GEOFFREY BEARD

Stucco and Decorative Plasterwork in Europe

Icon Editions

1817

HARPER & ROW, PUBLISHERS, New York
Cambridge, Philadelphia, San Francisco,
London, Mexico City, São Paulo, Sydney

STUCCO AND DECORATIVE PLASTERWORK IN EUROPE.
 For information address Harper & Row, Publishers, Inc., 10 East 53rd Street, New York, N.Y. 10022.

FIRST U.S. EDITION

ISBN: 0-06-430383-7

LIBRARY OF CONGRESS CATALOG CARD NUMBER: 82-49006

CONTENTS

For
PETER MURRAY
belatedly
on the occasion of
his sixtieth birthday

FOREWORD

In this prefatory note I want to explain the scheme of this book, and also say a little about the reasons for undertaking it. After an opening discussion of the stuccoist and plasterer and his medium, the text is divided into four sections, each of which documents, chronologically and by country, the contributions designers of stucco, stuccoists and (to a lesser extent) plasterers, have made to their art throughout Europe. In so wide a survey, inevitably personal selection and emphasis have arisen. The select dictionary of stuccoists and plasterers in particular contains a fraction of the many thousands who practised the art. The select bibliography lists not only the books and articles I have found useful, but a much wider range of titles of interest to specialist readers.

My researches into the history of plasterwork and stucco began in the early 1950s, and resulted in my *Decorative Plasterwork in Great Britain* (1975) Professor Peter Murray led me to consider the possibilities of further researches into 'Stucco in Europe' and introduced me to several Italian and Swiss scholars. It therefore gives me particular pleasure that he has accepted the dedication of this, the first book devoted to a subject of great interest to him. I have of course regretted on many occasions that the knowledge of stucco possessed by certain German scholars, and in England by Mr Alastair Laing and Dr Peter Cannon-Brookes, has either not been set out, or is confined to articles on specific aspects.

In several practical ways I am indebted to so many people (particularly parish priests and archivists), that I must thank them collectively, and thus inadequately. I should not have found much in the Swiss Canton of Ticino without the guidance and help of Professor Giuseppe Martinola, who, as the Bibliography shows, is a distinguished student of the past achievements of his countrymen. Professor Sir Ellis Waterhouse and the late Anthony M. Clark gave me the benefit of their extensive knowledge of the contents and decoration of houses and churches in Rome. Mr Brinsley Ford allowed me to search his voluminous indexes of eighteenth-century travellers to Rome. The late Dr Margaret Whinney discussed engraved sources with me. In Germany Eduard Kneitz and Erwin Emmerling made the resources of the stucco restorers Anton Fuchs of Würzburg freely available to me. Much help was also given on trips to Germany, Austria and Czechoslovakia by Alfons and Karin Popp, Edith Kneitz, Dr Rudy Wackernagel, Father Gebhard Spahr OSB, Dr Bruno Thomas, Dr Oldrich J. Blažiček, and the staffs of the Munich Stadtmuseum, Bayerisches National Museum, Art History Institute of Munich University, Bayerisches Landesamt für Denkmalpflege, and the National Gallery in Prague. Assistance with the expenses of travel and research was given generously by the British Academy, the British Council, the University of Lancaster and the German Academic Exchange Council.

Finally, help beyond the answering of routine questions was given by Professor Malcolm Campbell, Eileen Carvell, Sir Nicholas and Lady Judith Goodison, John and Helena Hayward, Margaret Knapp, York Langenstein, Pat Mueller, Dr F. S. Stych, Joan Wardman, and the directors and staff of my publishers. At the conspicuous end I am concerned to thank my wife Margaret, and daughter Helen, who endured my preoccupied presence and frequent absence with great cheerfulness, and in addition helped with objective criticism and unstinted encouragement.

University of Lancaster
February 1982

GEOFREY BEARD

Fig. 1. James Gibbs, unexecuted design for ceiling in St Martin-in-the-Fields, London, *c.* 1725.

I

MEN, METHODS AND MATERIALS

That an art so extremely interesting and widespread as that of the stuccoist should have escaped record and consideration in any collected form is to be regretted.

The basic materials used, gypsum, sand and water, albeit with additives, were easy to acquire at a reasonable cost. The use of stucco therefore ranged from simple protective coverings for structures, providing the base for wall-paintings, to fine modelling of decorative motifs, and sculptures in relief and free-standing forms. It is this capacity of the material to lend itself to a decorative function which is examined in this book. Stucco was modelled in the ancient world and throughout medieval Europe and Asia with little variation in material or technique. It achieved its most effective visual expression during the Renaissance, and was worked at high level of skill and iconographic complexity in the periods of Baroque art and the Rococo.

One of the first questions asked by those who are aware of the decorative functions of plaster and stucco is the difference between them. Plaster, they assume, is an English phenomenon; stucco is something similar but belonging to Continental Europe. The differences, however, are in the additions to basic chemical composition, and in the techniques of strengthening.

Apart from the isolated instance of the moulded stucco panels created at Nonsuch Palace by Nicholas Bellin of Modena for King Henry VIII, stucco was virtually unknown in England until the early eighteenth century. Then it was introduced by itinerant stuccoists to compete against the traditional English material of plaster. The stucco contained marble dust, and was fashioned around wood or metal armatures. The plaster contained lime, sand and water, but no marble dust. Its 'armature' or support took the simpler form of added chopped animal-hair. This gave enormous tensile strength in comparison with the more brittle stucco. A further distinction between the two was in the application: English plaster was applied in three thin, successive coats to a ceiling-construction composed of wooden laths nailed to the timber joists. The laths were nailed so that a slight gap was left between them. As the plaster was smoothed over them some passed between the laths and, on drying, keyed itself to the structure. In comparison, stucco was frequently applied direct to the brick surface of the roof vaulting, with no space left between the ceiling and the outer roof. While systems using timber laths were in use in Baroque Europe – at the Bavarian church of Die Wies, for instance – stucco acquired its independent aesthetic status by following more closely than in England the outline of the architectural structure itself. The undersurfaces of ceilings in Italy and Germany in the Baroque period were often brick or stone vaults, sometimes lined with terracotta tiles. This construction, for which there was much ancient precedent, was determined by the span to be covered and its height. While smaller rooms such as bed-

XII, XIII

and dressing-rooms could be erected in lath and stucco, staircase wells and important state rooms were given a brick structure.

While there is considerable literary evidence of the use of stucco in the classical world – Vitruvius and Pliny both describe methods of obtaining good stucco mixtures – its early use died away, and it was only 'rediscovered' in the early years of the sixteenth century. The chronicler Giorgio Vasari described in his *Lives* (1550) how Giovanni da Udine, a pupil of Raphael, was digging in the ruins of the Palace of Titus when he discovered certain rooms, completely buried under the ground, full of figures and scenes in low-relief stucco. He succeeded, after much study and many experiments, in achieving the ancient stucco formula. The most significant additive he stumbled on required 'chips of the whitest marble that could be found to be pounded and reduced to a fine powder, and then sifted'. This substance he added to the crystalline limestone (travertine) and well-washed river-sand and water. 'He had succeeded', wrote Vasari, 'without any doubt in making the true stucco of the ancients.'

With almost unvaried composition, this continued as the basic formula for all later stucco in Europe. The developments in technique, the creation of teams of workers with great skill in modelling, and advent of the sophisticated architectural settings of the Baroque and Rococo periods allowed the medium its full expression. The following chapters document the use of stucco in many incomparable settings, and show the virtuosity of which stuccoists were frequently capable.

The natural materials

Gypsum, one of the raw materials used for stucco and plaster, is the natural form of calcium sulphate. Distributed widely and plentifully in nature, it is mined from the earth as a snow-white soluble substance. Deposits in France under Montmartre in the nineteenth century led to the appellation 'plaster of Paris', and gypsum from Paris was sent all over Europe. First, however, the powder mineral was rendered more usable by being heated to about 150 degrees Centigrade to drive off the greater part of the water, and then further heated to some 400 degrees Centigrade, leading to the formation of the anhydrous calcium sulphate.

The other principal source of stucco is derived from limestone, a rock composed of at least 50 per cent calcium carbonate ($CaCO_3$). Man's use of lime as a cementing and plastering material is probably almost as old as his use of fire. Lime is, however, too caustic for use in plaster until it has been burned, when it becomes 'quick lime' or calcium oxide (CaO). It is then 'slaked' by being mixed with water, when it becomes calcium hydroxide ($CaOH_2$). Lime stucco was used for coarse first layers, and gypsum for the final layer.

Ancient Stucco

The Materials

Vitruvius, who was active 46–30 BC, indicated in his *Ten Books of Architecture*[1] the preparation of the various building-mortars. The lime needed to be well slaked and of a good grade, and the sand with which it was mixed, sharp and non-staining. Vitruvius noted that if proper attention had been paid to the slaking, the lime would stick to an inserted hoe (*ascia*) like glue, 'proving that it is completely tempered'. Roman law forbade the use of lime unless it had been kept for three years, by which time the gradual absorption of moisture from the atmosphere would have advanced the slaking process.

After the lime had been slaked carefully the sand or *pozzolana* could be added. 'Pozzolana' was a variant form of sand used for certain mortars, being volcanic in origin, and largely composed of alumina and silica. Vitruvius preferred what he called 'pitsand' – and of this, 'the best will be found to be that which crackles when rubbed in the hand, while that which has much dirt in it will not be sharp enough. Again: throw some sand upon a white garment and then shake it out: if the garment is not soiled and no dirt adheres to it, the sand is suitable.' The function of the sand was to separate the particles of slaked lime, so that the carbon dioxide in the atmosphere could work on the mix, and resolve the calcium hydroxide into calcium carbonate, with the inclusion of the sand and water. When these stages had been gone through, the mixture was ready to be applied to the wall.

The Layers

The surface for stucco decoration was built up in layers – first a very rough rendering-coat over the brick, and then at least three layers of sand-mortar, smoothed. The better the foundation of sand-mortar, the stronger and more durable would be the stucco. A mix of gypsum, lime and sand was next prepared, to which marble dust (itself a crystalline form of limestone) was added. After this had been worked with water and spread on, a second coat of medium thickness was applied, and this was well rubbed down. The final, seventh, thin coat provided a hard and dazzling white surface, and being set over a good foundation, did not crack. It also took colour well, if this was applied at the time it was polished, using a polishing tool known as a *liaculum*. As Vitruvius wrote: 'so, when the stucco on the walls is made as described above, it will have strength and brilliancy, and an excellence that will last to a great age'.

Each successive coat required a progressively finer grade of marble dust, whether for wall or ceiling, in order to achieve the 'strength and brilliancy' Vitruvius advocated.

For the construction of ceilings a frame was needed. This was made of wood interwoven with reeds. Later the reeds were replaced by laths – thin strips of wood were nailed with hand-made iron nails to the timbers. On the top of the frame a 'plaster' of lime and sand was spread. This received a rendering-coat (*trullissatio*), a layer of sand-mortar (*harena*), and finally a coat of stucco (*creta aut marmor*). The surface was then ready for decoration in relief or for painting, not needing the seven coats specified for a wall.

Vitruvius' instructions for seven coats – a rendering coat, three layers of *harenato* and three layers of *marmorato* – seem to have been generally followed in his time, but decline set in. Roman builders[2] of the Augustan period made do with less, and by the sixteenth to eighteenth centuries, four or five coats were common.

Renaissance stucco recipes

Renaissance theorists are as explicit as Vitruvius about the composition and preparation of the mixture for stucco,[3] although much of what they wrote was based on his earlier writings. A variety of basic materials and mixing ratios are listed, but north of the Alps, the finest results were obtained by a blending of earlier recipes.

Baroque stucco recipes

The processes of preparation of stucco then current are described by a number of eighteenth-century theorists such as Croker and Zedlers, but the most convenient account is that of P. N. Sprengel (1772):

> The ingredients of the stucco, as previously shown, are gypsum, lime (or chalk), and sand . . . Every stucco-worker mixes the ingredients as he, taught by experience, sees fit to. He changes sand, gypsum and lime, blended with water, into a paste . . . However the quality of the gypsum of binding fast, which hinders the artist from elaborating on his work of art, makes it necessary for the stucco-worker to add glue-water, for the glue-water retards the binding of the gypsum.[4]

Besides the glue-water mentioned by Sprengel, other retarding agents were used, such as milk, fermented grape-juice, curd, beer, wine, sugar and marshmallow-root powder. Wine in particular was used in the stuccoes created in the churches at Rheinau, Einsiedeln, Wilhering (in Austria), the Jesuitenkirche at Lucerne, and elsewhere. At Einsiedeln, over the three-year period 1724 to 1726, fifty-seven buckets of wine were so used.[5]

As well as slowing down the setting-time of the stucco, some of the agents were useful, too, in giving pliancy – glue, almond oil and curd had this effect. The pliancy and setting-time was also affected by the purity of the water used to mix the gypsum, the temperature of added ingredients, the humidity and temperature at the site, and the duration of the stirring of the viscous mix. Some stuccoists asked for fires to be lighted in rooms in advance of their visits in order to create an atmosphere in which the stucco would work with more pliancy, and cover a greater area.

In determining the proportions, there were two main factors a stuccoist had to learn to judge by experience: the quantity of water, determining the density of the gypsum, and the proportions of the materials added to give hardness and durability. Limewash and glue-water were useful for increasing hardness, but the principal method was to add alum, one of a series of 'double phosphates' of potassium and aluminium.

Each workshop had its variations on the basic formula, relying on its experience to achieve the maximum workability and adhesion to the framework of armatures. Usually, however, the slaked lime and sand were mixed in equal proportions. The lime-mortar formed by this mixture was kept fresh in tubs of water, until this pasty substance was mixed with further slaked lime (in the proportion 2:1), and with the additives described above.

Armatures

Vasari in his *Lives* indicated that if work in bold relief were required, the stuccoist should fix nails or suitable framed supports – armatures – to hold the weight. When the job was a large low-relief he advocated that nails should be hammered-in to leave the height required projecting. This support would hold the first coat.

In German Baroque decoration the first layers of plaster were reinforced with various kinds of animal-hair, in particular that of the deer (*kalberhaar*), or with hemp, straw or reeds. Wooden pegs or metal nails were used, as well as iron-wire in rod and mesh form. The metal parts, on oxidizing within the wet mass, produced a stiffening-agent, but care was needed to see that the wrapping of the metal prevented rust from leaching through to stain the modelled surface.

My conversations with stucco-restorers in Lugano and Würzburg in 1966 and 1979–80 confirmed that they still practised eighteenth-century techniques. The wood they used for frames was usually willow, and wood was more favoured than iron, because although, as noted, the rust from the iron 'stiffened' the mix, it was also capable of splitting the stucco. The wood or iron armatures were wrapped with canvas thread. A rare Renaissance example of the use of animal-bone armatures was noted by Professor Irving Lavin undertaking restoration in Florence in 1968. Presumably bones recommended themselves by being capable of resisting the slightly caustic stucco-mix, or the generation of heat in slaking.

To support a figure, a large wooden or metal armature was anchored at a steep angle to the wall and then set into the back of the figure. The legs and arms were supported by pieces of wood about half-an-inch (12 mm) thick. The fingers and the small objects carried by figures were set on wires – even violin strings, suitably stiffened with gypsum, have been noted – or on pieces of strip-lead, leather or twigs. An instance of metal and wooden supports set into the wall (here to support the great pulpit) was to be observed during the 1978–9 restorations at the abbey church at Melk in Austria. Such methods of support were universal in Baroque Europe.

Moulds

When Vasari wrote about the use of moulds in the sixteenth century he described a method that was to remain common throughout the following three centuries. Ornamental mouldings were made by means of reverse-carved wood moulds, 'and some stucco which is half set is beaten into the mould which has first been dusted with powdered marble. The stucco is beaten heavily until the ornament has been transferred to it when it is taken out and cleaned up . . .' I have examined and compared a number of accounts, and made observations while this process was being undertaken. The stages were as follows.

The enrichment was modelled in plaster or clay, and a mould was cast from the model in plaster. This mould was soaped, or coated with some preparation to prevent adhesion of the plaster. A semi-liquid plaster was poured into the mould, and the mould was immediately applied to the ceiling and held there by a stay from the scaffolding until the plaster was sufficiently set to allow its removal. The enrichment was afterwards cleaned and touched up where necessary using modelling tools. Where a vast number of *putti* or other small figures were needed, the heads and arms would be moulded, and then hair and other distinguishing features modelled as added stucco-layers.

Two variations in technique employed by the stuccoist were 'hollow moulding' and 'press stucco'. P. N. Sprengel, writing in 1772, describes the advantages of moulding ornaments in advance of need, and of hollow moulding:

> Frequently, the stuccoist is forced to speed up his work, and perhaps even more frequently he gets a small payment. Both give reason enough to mould a store of heads, masks and even flowers of plaster in the leisurely hours of winter, in a mould, which he, like the sculptor, has made of plaster. He models the ornament for which he is going to make a mould in its full size in clay or stucco, and over the completed model he pours the gypsum to form a mould, with or without centre-pieces, whichever the shape of his model requires. Occasionally the artist asks the sculptor to make him a flower in gypsum according to that mould. He can save gypsum if he moulds the ornaments hollow. In the place where the ornament is to be fixed he puts several big-headed nails into the wall. He drills a hole for each nail into the ornament of gypsum, and when the moulded gypsum-plaster is solid he puts the ornament on the nails and fixes them with a mixture of one third lime and two thirds gypsum. Since this mixture is made mainly from plaster it binds within a short time.[6]

The main advantage of hollow moulding, however, was the reduction in weight of figures.

In 'press stucco', 'press moulds' of hardwood were held against the stucco *in situ* while it was still soft. The technique has been noticed in ceilings of the late sixteenth century in the Thuringia-Saxony area, and was obviously a speedy method of stamping often-used motifs such as egg-and-tongue moulding, and the various forms of leaves in garlands and ribands. The background to which this ornament was applied was generally slightly recessed below the main surface of the ceiling, framing the ornamentation and forming a part of the decorative whole.

Fig. 2. Wooden mould for a Rococo swirl by Ignaz Günther (1724–75), in the Stadtmuseum, Munich.

Press-moulding could also be used to provide a coloured under-layer, revealed in part by cutting through the upper layer with a pointed tool. Two moulds were used, identical in pattern, but one smaller than the other. The smaller mould was used to first-stamp the pattern. A thin coating of stucco was then applied, and the larger mould was used to stamp the pattern again. After part-drying, undercutting through to the base-pattern was undertaken.

Diminished mouldings, such as those running upwards on the curvature of a dome or cupola, were achieved by the use of special hinged moulds that allowed the profile to be altered.

Models for moulds were often commissioned from specialists in modelling. The building records of the Hofkapelle of the Munich Residenz indicate that in 1630 the artist Matthias Schreine had 'cut models for the Kurfurstliche Hofkapelle which will be of use elsewhere as well'.[7] When the Jesuit church in Innsbruck was stuccoed in 1634–7, models were supplied by a Tyrolean sculptor, F. Nut.[8] Many stuccoists, of course, were adept at making their own moulds.

The frequent re-use of moulds – which in the sixteenth and seventeenth centuries were particularly expensive – made some identification of the stuccoist a possibility. Known through documentation to have worked at one house, he might be suspected at another by the use of the same moulds. This occurs frequently in Scotland; and even without the plentiful documentation, it is possible to identify work by the Wessobrunn school in Bavaria and elsewhere by their common moulds and repetitions of the same engraved source.

For a mould intended to last and give repeated good impressions boxwood was often used, but examples have been found in yew-wood, various close-grained hardwoods, soft blackstone, iron (particularly in Scotland), as well as *gesso* (gypsum and size), wax and gelatine. Following the use of wax moulds in the later eighteenth century, gelatine moulds were introduced in the 1840s in France by H. Vincent. Their flexibility was useful in casting statues and plaques, and several pieces produced by this method were shown in London at the Great Exhibition of 1851. The only drawback of gelatine moulds was that unless they were well cared for, they became hard and brittle. The gelatine was, however, obtained easily from the bones, hides and hooves of animals, and could be melted down and re-used, so that fresh moulds could be created after a few jobs.

When the Ticino stuccoist Giuseppe Artari was in England at Trentham in 1737 he made use of the Broseley pipe-clay to make his moulds.[9] It had obviously suggested itself because of its suitability for reproducing fine details, and was rendered 'permanent' by firing the clay mould itself.

Collections of moulds survive in the Geffrye Museum, London; and in the Würzburg Residenz there are about twenty examples from those used by Antonio Bossi in the 1750s. Two or three moulds survive in the Ignaz Günther Haus (part of the Stadtmuseum) in Munich, including a fine example for creating a Rococo swirl.

Fig. 2

Tools

While the principal tools used in ancient Egypt were practically identical with those used today, the variations occurring over a long period necessarily prevent description of a 'standard set'. We may start in England with the description of the tools given in Joseph Moxon's *Mechanick Exercises* (1703). Moxon lists ten main tools in use by stuccoists. *Fig. 3* However his illustration, reproduced here, shows many tools used by bricklayers and tilers rather than plasterers. Similarly an illustration from P. N. Sprengel's book published in Berlin in *Fig. 4* 1772 shows a few stucco-tools among those of the stonemason.

In addition to illustrating the stuccoist's tools, Moxon continues:

> . . . there are some things yet remaining, which tho' they cannot be properly called Tools, yet they are utensils, without which they cannot well perform their work.

He then lists ladders, fir-poles for scaffolding, and putlogs, or seven-foot (2 m) pieces of timber which lie from the scaffolding poles into the brickwork and support the ten-foot (3 m) fir boards, of one- or two-inch (2.5 or 5 cm) thickness. The cords which tied these various poles and planks together were to be 'well pitched to preserve them from the weather, and rotting'. There were also sieves of several sorts to 'sift the Lime and Sand withal', and a 'loame-hook, beater, shovel, Pick Ax, Basket and Hod, which commonly belong to Bricklayers . . .'.

The eighteenth-century stuccoist had some tools which were additional to or replaced items in Moxon's list. He used a tub, a gypsum-trough, a gypsum-pan and a gypsum-bowl. The stucco-dough was scooped up with a mason's trowel, often on to a 'stool' with a circular top which could be raised on a threaded stem or by a dowel inserted into higher or lower holes (item XII). From

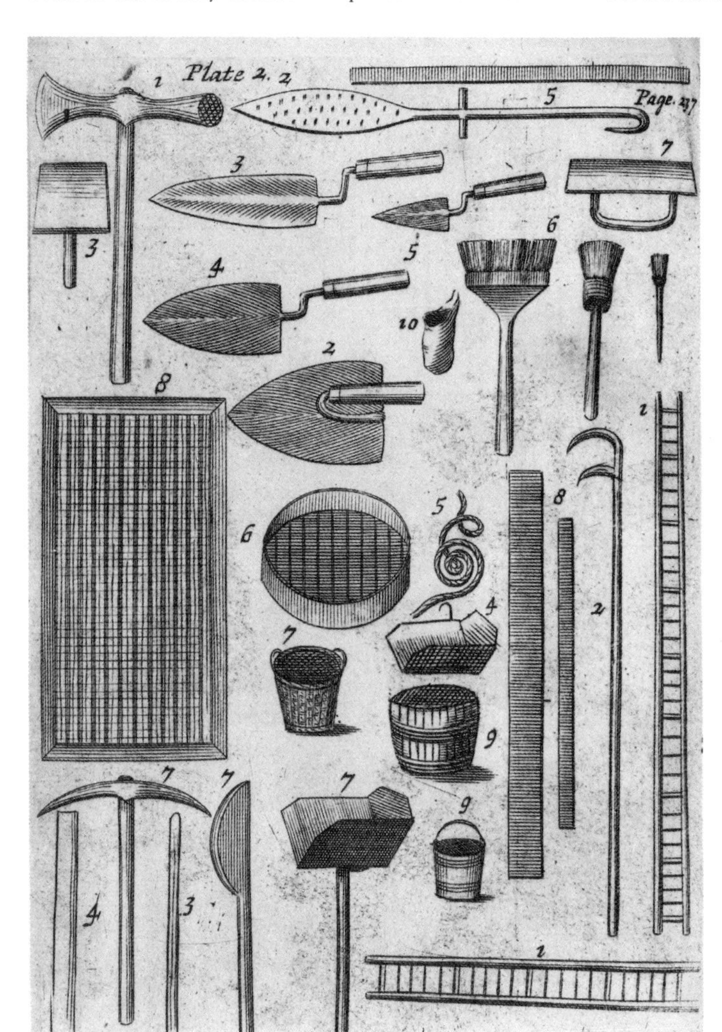

Fig. 3. Plasterer's tools from Joseph Moxon's *Mechanick Exercises* (1703).

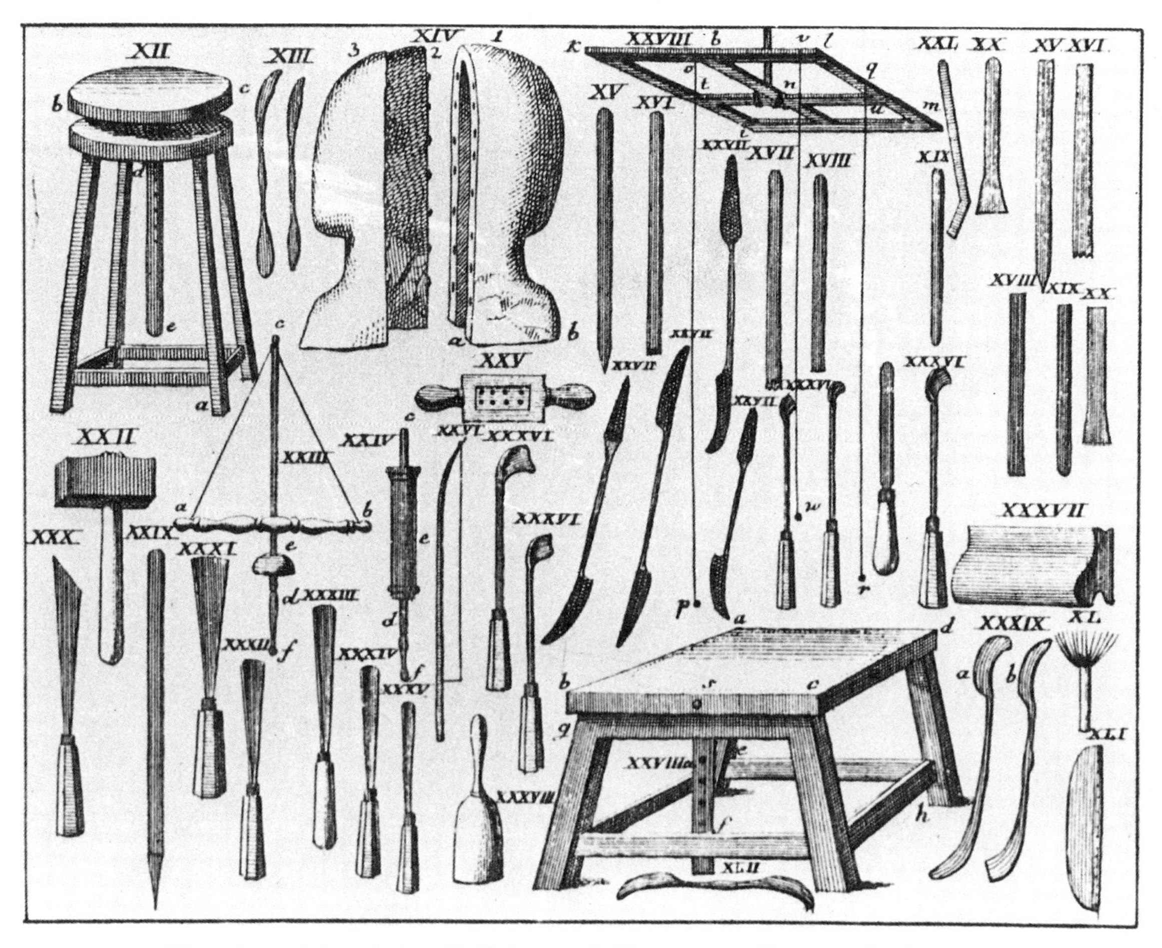

Fig. 4. Stuccoist's tools from P. N. Sprengel's *Handwerke und Künste . . .*, Berlin 1772.

this surface the stucco-dough was applied with a gypsum-trowel. Various spatulas or poussir-irons were used to apply and even-out the gypsum, and modelling-sticks served to bring the mixture into the correct shape, or to press, spread or turn it. The poussir-iron (no. XLII in Sprengel's illustration) was a double spatula. Some irons were straight, others raised, some were round or hollow on one side. With the raised iron the stuccoist scooped out the indentations. He cut the hard stucco with other irons having fine saw-teeth. The gypsum-knife and a series of scrapers also assisted in the shaping processes, and were held so that they cut off tiny gypsum flakes. A coarse paintbrush served for removing the dust and bits from the work, and fine soft paintbrushes were used for whitewashing with a lime-wash over the completed parts of the design.

The diminishing of columns, a complicated and difficult operation which required great care to avoid exaggerating the entasis, was accomplished by the use of a special tool, a 'floating rule' known as a trammel. 'Floating' merely implied the process of putting on the stucco and checking its levels with a long rule.

Methods of work

There is little documentary evidence to establish the precise methods of work used in the past, but from observation of surviving stucco, and conversations with stuccoists and restorers, some facts can be pieced together.

It was essential for plasterers and stuccoists to work from scaffolding or trestles. There is no evidence at any period of their working from cradles slung beneath a ceiling, although the architect James Wyatt is said to have ventured something akin to this when in St Peter's in Rome (*c.* 1764) – 'lying on his back on a ladder slung horizontally without cradle or side-rail, over a frightful void of three hundred feet'.[10] It was more usual to work from scaffolding, or planks placed across two-legged 'great trussels', set within six to seven feet (1.5–2 m) of the surface to be stuccoed.

The considerations proposed in the 1660s by Sir Roger Pratt (and in the eighteenth century by James Gibbs) were those which faced most architects and stuccoists in planning a ceiling design. Writing about Italian ceilings, Pratt listed,[11] 'of the several frets and divisions for ceilings: *1* Breadth of the soffit; *2* The ornament of it; *3* The form of the panels; *4* The beautifying of them; *5* Design; *6* Colouring; *7* Panels and Plaster.'

The sequence of work went something like this. The ceiling was measured, and the centre axes and panel positions were marked with a charcoal pencil. The rough disposition of wall-decoration comprising bas-relief heads and swags was found so marked in charcoal in an English house, Honington Hall, Warwickshire,[12] when it was restored in the mid-1970s, necessitating stripping back to the brick and thus revealing a preliminary stage which the stuccoist assumed would never be revealed. The cornice was then set, usually over a timber framework secured to wall and ceiling. This cornice was either built up from a profile moulding or made up of separate component parts, such as modillion, ovolo, dentil, cyma reversa and astragal.

Fig. 5. Moulded head. Life-size stucco figure of a Roman slave, on the Hall ceiling of Moor Park, Hertfordshire, *c.* 1730, attributed to Giovanni Bagutti.

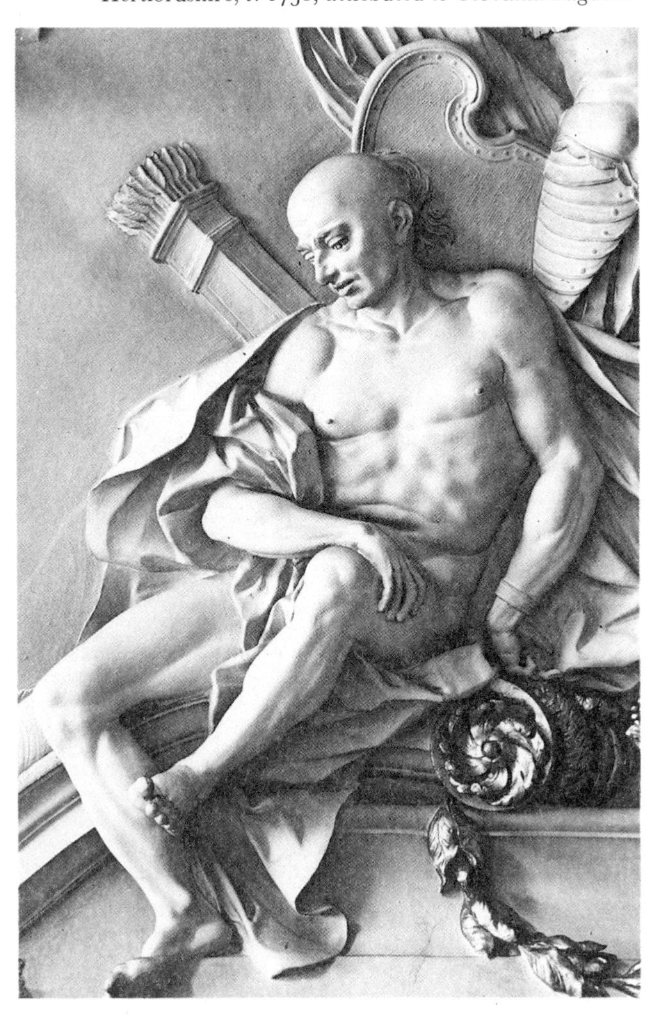

The panel or rib mouldings were then 'planted' in position after being 'run' on a bench or running-board, and cut to the proper lengths to mitre. A variation was to run the ribs *in situ* by pressing a mould (a cut-out profile of the shape required) against the soft stucco which had been laid on the ceiling along the measuring marks. Mouldings so formed were stronger than those 'planted on'. Planting was used where there was little space for working the running mould, as on short or circular pieces, or within small, intricate panels.

Whether run *in situ* or turned out of moulds on the bench and carried up to the ceiling for planting, the stucco needed to be well keyed to the cross-hatched surface, and for planting, adhesion was aided by the use (as in pottery) of a 'slip' or thin cream of stucco applied to the base of the ornament to be applied. The planting of large circular ribs was achieved by making the circumference in four segments on the bench: small circles or semi-circles could be run *in situ* on the ceiling by attaching a mould of the desired profile to a radius arm, pinned so that it inscribed a true arc. Direct modelling *in situ* minimized labour and cost, since pieces made on the bench had to be first modelled and then fixed. Running on the bench was indicated where there was a considerable amount of repetitive ornament, or, as noted, when working in restricted areas.

The ornamentation frequently found within the under-surface or soffit of the rib had each leaf or fruit modelled on a separate armature. The motif was then intertwined with its neighbour, and the decoration built up to form a satisfactory pattern. The detail was sharpened with modelling tools. This form of naturalistic decoration was at a high level of accomplishment by the 1660s. There are splendid examples in the London City churches designed by Sir Christopher Wren, at English houses such as Sudbury Hall and Belton House, and in German churches stuccoed, in

68

Fig. 6. Two winged cherubs' heads after war-damage, showing the layers of stucco, *c.* 1740, Würzburg Residenz.

particular, by Giovanni Battista Carlone and the Lucchese family. Outstanding are the churches at Passau, Waldsassen, Gartleberg, Amberg and Speinshart.[13]

47, 53
v

Stucco figures and medallion portraits

Standing and reclining figures, the range of smaller angels, *putti*, caryatid figures, conjoined heads, animal-forms and medallion portraits of mythological or actual personages make up the larger ornaments fashioned most frequently in stucco. The basic construction of all stucco figures was similar, and involved varieties of hollow moulding, support on metal or wood armatures, and filling with various substances. The construction of the impressive caryatid figures by Giuseppe Artari (*c.* 1748) on the staircase at Augustusburg Castle, Brühl, was established in post-war restoration, and confirms what I have been able to see elsewhere. The diagram is based on the cross-section of a figure given by Wilfred Hansmann in his account of the Brühl staircase.[14] The filling of charcoal, or in the case of small *putti*, of straw, allowed the figure to have 'body' without the hollow cast collapsing, and was useful in maintaining an overall low weight. While the armature was usually placed centrally in the figure to support its height, subsidiary armatures were required where croziers, staffs, spears or similar objects were held at an angle. A most successful unobtrusive placing of two iron armatures is that of the figures in the mighty Asam high altar group of the Assumption of the Blessed Virgin, at Rohr (1723). One metal armature bound with straw is driven diagonally into the lower part of the group, another is set horizontally, at the level of the shoulders of the Virgin. Neither is visible to the beholder.[15]

VII

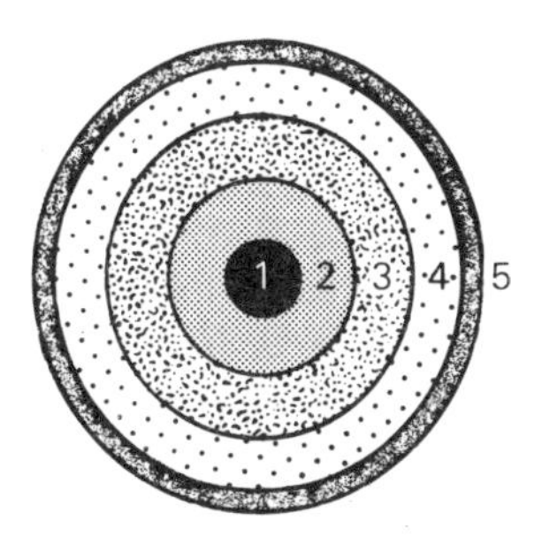

1 Iron 2 Chalk/powdered brick
3 Charcoal 4 Modelling clay 5 Fine stucco

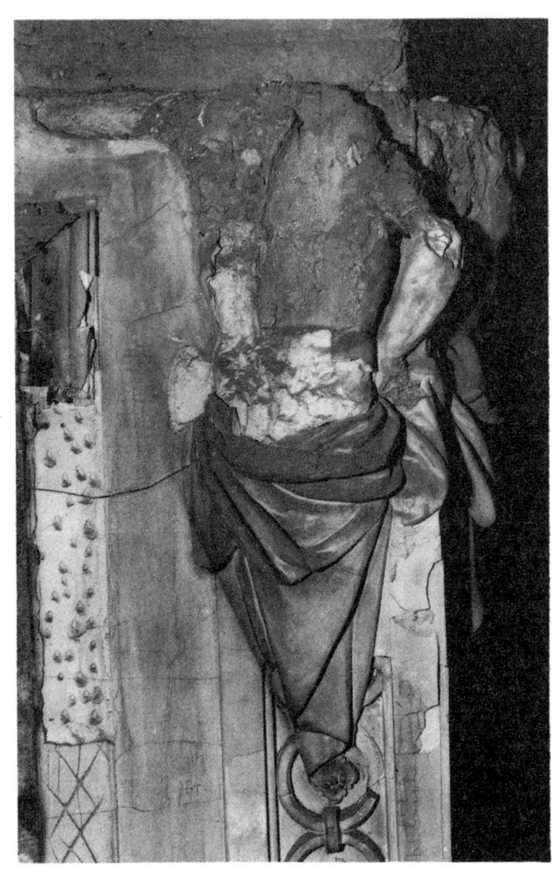

Fig. 7. War-damage to a stucco caryatid figure revealing nail armatures and showing scratch-marks for keying of layers, Würzburg Residenz.

It has been possible to examine the manner in which figures at Weltenburg and Straubing support their large metal circlets or crowns. In each case the metal of the crown enters the arm of the supporting *putti* as a spiked armature.

Many low-relief figures did not need the support of armatures. Supports such as nails could be used to key each layer. The absence of armatures is revealed in the photograph of a war-damaged caryatid figure at the Würzburg Residenz. *Fig. 7*

There is some indication that the making of figures was a specialist occupation, not practised by all stuccoists, although most were more than adept at their construction. This specialization is suggested by a comment of Lord Fitzwalter when Artari was working for him with Giovanni Bagutti at Moulsham, Essex (1731): 'my agreamt. was only with Mr Bagutti and Mr Artari who did the Bustos & Figures assisted him'.[16]

Confusion may arise between carved wood and stucco figures when the former are covered with a chalk layer, or are gilded or silvered. The only aids to detection are knowledge of documentation reinforced by a physical examination. The reverse of the figures usually indicates the material used, or wood-worm holes may provide evidence. Drapery folds on Baroque statues tend to be more angular and sharp in carved wood than in stucco, and wood figures were rarely set higher than the top crestings of pulpits, choir-stalls and side-altars. Carvers generally eschewed incised modelled clouds, which the stuccoist created partly in order to hide the armatures which supported figures on cornices, arches or flanking-altars. Stuccoists have simulated marble, and only close examination at positions where a moulded head is attached to the body, and touch (where this will be non-injurious to the surface) allow detection. Particularly difficult to distinguish are figures set at great height – where marble was, however, less likely to be used through consideration of weight. High positions in churches needed the lighter, hollow stucco figures, while standing figures on altars were usually of wood.

Medallion portraits were almost invariably moulded, and then likenesses or additions were provided by modelling *in situ*. They are encountered from Renaissance times through to the nineteenth century. The most common subjects were the heads of Roman emperors, although some profile portraits have been noted, and depictions of philosophers and literary figures were common.

Colour

Colour is a complex study, made more so by the changes which have occurred in existing schemes due to age, atmospheric effects or restoration. On the last, Gottfried Huth wrote in the eighteenth century that 'the arts have no greater enemy than fashion'; and even when fidelity was intended, it was not necessarily achieved.

Five groups can be set out as the typological pattern:[17] *a* white or grey stucco on a wall of the same colour; *b* white stucco on a polychrome wall; *c* polychrome stucco on a white wall; *d* polychrome stucco on a polychrome wall; *e* metal applications (gold, silver, copper in foils) to stucco and/or wall.

What past ages understood best of all in stucco was the use of 'white-on-white'. The revival in a serious way of white-on-white stucco may be traced to Rome in the early 1630s. The great figures of the Baroque, Francesco Borromini and

Pietro da Cortona, made early and impressive use of monochrome stucco.

White-on-white stucco was usual in the British Isles in the seventeenth century and the first half of the eighteenth century. It was also adopted in southern Germany, by the Italian artists working in Munich at the Theatinerkirche, [50] or the Wessobrunn Schmuzer family at Obermarchtal, [55] Friedrichshafen and elsewhere at the end of the seventeenth and start of the eighteenth century.

The use of the colour-schemes *b* to *e* can be found across Europe, particularly in the eighteenth century. While to the casual observer such colour-schemes may seem to lack any system or pattern, liturgical requirements instilled order into some of them. Colour, which French theorists of the 1670s such as Roger Piles[18] asserted was superior to drawing, was used most intensely in the upper areas, where it surrounded bright frescoes which enriched the whole structure. As for the colours themselves, red was associated with blood as a symbol of sacrifice and martyrdom, and black with mourning. Green extended hope, but yellow suggested envy and jealousy. With white for purity and holiness, blue for humility (and therefore used for the robes of the Virgin), and purple, brown and grey for penitence, a spectrum of colours was available to add to a wide variety of attributes and symbols of Apostles and saints.[19]

Occasionally pigment was worked into the stucco mass, as by Peter Anton Moosbrugger (1732–1806) at Herlsau in Appenzell in 1792. Moosbrugger used shades of grey and ochre.

In recent years a number of restorations have focused attention on the question of the colours originally used. Writing in 1973, A. F. A. Morel set out the problems for the restorer as follows: the different composition and method of manufacture of modern paints; the changes in the original colour-tones; the paucity of documented information about the colours used. In autumn 1970 he noted that cleaning of the nave of Katharinenkirche at Laufen uncovered proof of the 1755 colour-schemes by the Moosbruggers.[20] This permitted an exact reconstruction of the original setting, which was kept to soft *Régence* shades. For the restoration at the great church of St Gallen in Switzerland in the 1960s,[21] careful analyses and paint-scrapes were made, and the colours discovered were confirmed by nineteenth-century accounts. Nevertheless, the final painting, in copper-green and beige, has been criticized as 'new' and lacking in any reference to the past. At the restoration of Peter Anton Moosbrugger's work at Bernhardzell in 1955–6, the redecoration was kept to the original soft grey with green markings, and a yellow-and-pink background, an example of polychrome stucco on a polychrome background.

Most nineteenth-century restorations can be detected by heavy tonality, while those of more recent date have often provided schemes of doubtful fidelity to first intentions but considerable attraction. Few will quarrel with the bright colours at Steinhausen, or the delicate pink-and-white, with heavy gilding, at Ettal.

Gilding

Gold beaten into extremely thin sheets known as 'leaf gold' was applied to stucco, with the use of an adhesive called a 'mordant' – usually dried linseed oil, and by one of the two processes, oil or water gilding. Oil gilding was simple and durable. Its drawback was that, unlike water gilding (in which a pigment was applied in solution over the leaf gold), it could not be burnished. Both, however, were used on stucco, with a preference for oil gilding as this avoided softening of the surface by water. An important aspect of gilded decoration was incising the gold leaf with sharp tools to produce a luminously patterned surface.

Gilding was in evidence from ancient times, throughout Egypt and classical Greece and Rome. It was a major part of architectural enrichment in Byzantine times, and was in use in France in the late seventeenth century, although there is no evidence for an exact date for its introduction to European Baroque decoration. In Germany it was used on two early schemes – the Hofkapelle at Ludwigsburg, stuccoed about 1718 by Diego Carlone to designs by its architect, Donato Giuseppe Frisoni (1683–1735), and in the Pagodenburg at Nymphenburg (1716–19) for the Elector, Max Emanuel.

The skills of the gilder – a specialist application involving knowledge of stucco composition, were akin to those required for the simulation of marble by working in *scagliola*, and a variant, *stucco lucido*.

Scagliola (stuckmarmor)

Scagliola derives its name from the Italian word *scaglia* – scales or chips of marble. This coloured and polished stucco has been used from Egyptian times to imitate, and provide a cheaper alternative to marble. Its re-introduction into Italy in the early seventeenth century is credited to Guido Sassi (1584–1649), working at Carpi in the state of Modena, and possibly the same Lombardic mason who was titled Guido del Conte. An early seventeenth-century use of the material in Ger-

Fig. 8. Jan Vredeman de Vries, engraving of decorative panel, *c.* 1570.

many (*stuckmarmor*) is the pink and grey panels moved about 1725 from the Munich Residenz, and set into the small oratory of Schloss Schleissheim, a few miles to the north, by the stuccoist J. G. Baader (b. 1692). In England it seems to have been first used in work for the Duke and Duchess of Lauderdale on the elaborate chimneypiece in the Queen's Closet, Ham House, Surrey, *c.* 1674.

Scagliola was made usually from a crystalline form of calcium sulphate (gypsum). Pure gypsum was broken into small pieces and calcined. As soon as the largest fragments lost their brilliancy, the fire was withdrawn, the powder was passed through a fine sieve, and it was then ready to mix with sand and a glue (and later isinglass) solution. Into this paste the colours required to imitate the marble were diffused by chopping and rolling the mix into a thin pastry-like covering. The surface was then polished with pumice or various grades of scraper (later abrasive papers), using oil (or for ceilings, water) as a flux. The Würzburg firm of Anton Fuchs, who restored the *stuckmarmor* surfaces at the Residenz damaged in the Second World War, provide a modern example of the process. Variant recipes specified the inclusion of ground marble, spar or granite, and additives such as alum which made the *scagliola* hard enough to be laid as a 'marble' floor. The various surfaces created from *scagliola* could be incised with pattern-lines $\frac{3}{8}$ inch (5 mm) deep which were filled with contrasting colours to represent foliage, coats-of-arms and figures. *Scagliola* was a cheaper substitute for *pietre-dure*, the hard and semi-precious stones which were fashioned at Florence and other centres from the late sixteenth century.

One of the most frequent uses of *scagliola* or *stuckmarmor* was to form the surface of columns and pilasters. The column to be covered was usually constructed from nine pieces of wood in a grid pattern. On each of four sides, a solid piece of wood was nailed to the grid and planed on its outer surface to a correct curvature for the column. On to these surfaces was trowelled a thick layer of very coarse plaster, its rough surface acting as a key for the final thin layer, $\frac{1}{4}$ to $\frac{1}{2}$ inch (3–6 mm) thick of *scagliola*. This method of construction is confirmed by one 'L', writing in the *Somerset House Gazette* for 1824 (p. 381) of a visit to the Coade stone artificial stone manufactory in

Fig. 9. Jean Lepautre (1618–82), detail of an engraving of acanthus ornament, *c.* 1670.

London. He was greatly impressed by its *scagliola* department, commenting that their work equalled 'marble in brilliancy, smoothness and variety of tints'. We may observe the product for ourselves thanks to the fortunate survival of a column (which had been made too long and was therefore not erected) for the 1793 decoration of the Chapel of the Royal Naval College, Greenwich.[22]

The most impressive examples of *scagliola* work are altars and altar frontals. There are three at Weingarten, including the high altar designed by Frisoni and executed by a *stuckmarmorier*, Giacomo Antonio Corbellini (?1674–1742). In addition, the Zimmermann brothers used mottled green *stuckmarmor* columns to great effect at Buxheim. These were created for them in 1727 by another specialist, G. D. Weis, and rival those in delicate blue-grey and pink at their pilgrimage church of Die Wies. Dominikus Zimmermann also designed or executed several *stuckmarmor* altars at Landsberg, Steinhausen (worked in dark grey by Joachim Früholzer, 1747–50), and Buxheim.[23]

In the nineteenth century *scagliola* enjoyed a great vogue, and dictionaries of architecture and building magazines abound in details of its use.

Stucco lucido

A much rarer process than *scagliola*, allowing little control over the dispersal of colour, was known as *stucco lucido* or *stucco lustro*. A layer of fine mortar

Fig. 10. Drawing for portico ceiling by Giovanni Battista Ricci, *c.* 1612, for St Peter's, Rome (*see pl. 22*).

Fig. 11. Jean Berain, decorative panel, engraving from *Ornamens inventez par J. Berain*, 1711.

was damped and then coated with a thinner layer of fine gypsum. While this was wet, pigment colour was scattered across it, and 'trailed' with a pad, brush or fine-meshed comb. When nearly dry, the surface was ironed with a hot flat-iron and glazed to a shine not unlike that of calendared paper. The temperature of the working location was critical: while the setting of ordinary stucco could be delayed for a day by damping, *stucco lucido*, composed of neat gypsum and pigment, set rapidly. This made it a useful variation for undulating surfaces, where polishing by scraper was difficult. An example of the work is the fine twisted columns of the high altar at Osterhofen by Egid Quirin Asam (1731–2). Modern restoration of some ceilings at the Würzburg Residenz has recreated them in *stucco lucido*, polishing the stucco by the use of water and fine abrasive papers with several hours of rubbing.

Training: guilds and academies

To be competent enough to satisfy patrons, the stuccoist chiefly trained through an apprenticeship system regulated by trade guilds.

The English system[24] was well established and later allowed masters to take apprentices for instruction for periods of seven to eight years. During the time of an apprenticeship the master taught the trade and set an example of good workmanship. When the apprenticeship was finished the test-work set to the pupil was examined by representatives of the livery company governing the trade. Then, at three successive meetings, the apprentice was 'called'; if no one objected to his election he was sworn in as a member of his guild before the mayor at the borough court.

In Germany as elsewhere craft guilds took an important part in town-management from medieval times, but their prime function was to impart skills and regulate competition. No master had the right to monopolize the supply of raw materials, and he could not achieve an undue advantage by having limitless numbers of apprentices. Prices were fixed, and wages for journeymen determined. Most important, however, was the monopoly of producing their particular wares, or of practising a skill. No one practised without belonging to a guild, and in some cases this right was reinforced by a town charter. Before anyone could belong to a German guild his parentage was assessed. Newcomers to the town, sons of serfs and illegitimate children were usually excluded. The age for apprenticeship was twelve, or more rarely fourteen (as in England). The pupil served his time as apprentice, then worked as *Geselle*, or journeyman, over several years in different towns (the *Wanderzeit*), before he submitted a prescribed Masterpiece for the approval of guild officials. At the start of each of his three stages – apprentice, journeyman, master – the member paid a fee to the guild, and admission as a master implied to all that he was possessed of a certain capital.[25] There was a concern to master many skills, and an overriding necessity to master mythology and Christian iconography in order to satisfy demanding and knowledgeable patrons.

Patrons could provide at their courts an artistic freedom of a sort not tolerated by the guilds, and it

Fig. 12. Plate showing bandwork from Paul Decker's *Fürstlicher Baumeister, I*, Augsburg 1711.

was they who were to loosen the guilds' control. The final challenge to the German guilds came in the seventeenth and eighteenth centuries from the academies, and from the attempts of merchants 'to expand or change industrial production . . . and to increase profits'.[26] What survived most attempts at reform of the guilds was a laudable insistence on quality, though this itself was the greatest hindrance, in a manufacturing sense, to increased production.

The guilds had been pre-eminent in teaching about materials and methods of work, whereas the academics pursued classical disciplines. Academies in Paris (1648) and Rome (1666) set the early patterns of private instruction, and Germany emulated them in those at Nuremberg (1662) and in its official academy at Berlin (1699). Despite the ravages of war, a private academy was established in Vienna in 1692, and formed the climate for the foundation of the Vienna Academy of Fine Arts some thirty years later (1725). For the stucco and fresco arts, however, the most important German academy was that established at Augsburg in 1710.

In addition to formal apprenticeship or academic studies as a form of artistic training, there was the 'patrimony' system of the Ticino stuccoists, with father training sons (p. 57); and also work in the studio of a master craftsman. This was common in Renaissance times, but was replaced gradually by membership of academies, or by study-visits to an important city such as Rome or Paris.

Among the professional and antiquarian associations which flourished in Rome in the eighteenth century was a 'Universita degli Stuccatori o Plasticatori'. No records about this body appear in the Archivio di Stato, although it is possible that the lists of those enrolled will eventually be found in the voluminous notarial archives, since notaries usually took the minutes of the meetings of the various 'universities'. On the occasions the fellows (*soci*) – who were the majority of those enrolled – were listed if present. Stuccoists are absent, as far as can be established, from the indexes of the Accademia di S. Luca; nor do they appear as prize-winners in its academic competitions (1702–1904).[27] The paucity of recorded information about stuccoists encourages cautious reference to the other arts. The Scottish architect Robert Adam (1728–92) spoke for many European artists of his time, when writing from Rome that: 'I hope to have my ideas greatly enlarg'd and my taste formed upon the solid foundations of genuine antiquity . . . I am convinc'd that my conceptions

Fig. 13. Drawing by James Gibbs, *c.* 1730, for the ceiling of Gubbins, Hertfordshire. The ceiling was destroyed *c.* 1836.

of Architecture will become much more noble than I coud ever have attain'd by staying in Brittain . . .'.[28]

Sir Christopher Wren touched in 1694 upon the fundamental weakness of the English training through apprenticeship rather than in the academies of the Continent, in writing of the lack of 'education in that which is the ffoundation of all Meckanick Arts, a practice in designing or drawings, to which everybody in Italy, France and the Low Countries pretends to more or less'.[29] This handicap may, however, be exaggerated as far as stuccoists are concerned: there is no reason to think that English plasterers were less well trained than stuccoists of the German Wessobrunn school, although the second did have readier access to a wide repertory of well-realized engravings. Influences reached early eighteenth-century English plasterers via the few architects who had trained on the Continent and knew something of ancient stucco, such as Archer, Gibbs and Kent. Kent won the Pope's prize for drawing in 1713, and was also 'the first of the English Nation' to be admitted into the great Duke of Tuscany's Academy of Artists, 'which is an Honor to his Native County of Yorke'.[30]

Fig. 1

Engraved sources

Engravings of ornament provided architects and craftsmen with readily transported patterns. During the sixteenth century the work of Flemish engravers predominated, and their illustrated emblem books helped to diffuse Renaissance ornament throughout Europe. Their engraved biblical and mythological scenes also opened the way for the serious use of symbolism by architects and stuccoists. The *Emblemata* of Andrea Alciati was popular: first published in Italy in the sixteenth century, it ran into nineteen editions, and was translated into many languages. Cesare Ripa's *Iconologia* (Rome 1593 etc.) was also invaluable.

During the High Renaissance and Mannerist periods, the issue of many elaborate engravings spread throughout Europe a wide repertoire of ornamental motifs such as *putti* and grotesques, scroll- and strapwork. Caryatids, sphinxes and terms were popularized by the books and single engravings of Jacques Androuet du Cerceau I (active 1549–after 1584) and his rival Jan Vredeman de Vries. *Fig. 8*

From the early Baroque period, the cartouche, or shaped tablet in an ornamental frame, was popularized by Stefano della Bella in *Nouvelles inventions de cartouches* (Paris, 1646–7).

The many engravings issued by François Cuvilliés and others helped to introduce the Rococo style to South Germany. Cuvilliés published his *Livre de cartouches* in 1738, and followed it with elegant suites of many other subjects. *Fig. 14*

In Germany, two books were of outstanding significance. These were J. J. Biller's *Neues Zierathen Buch von Schlingen und Bändelwerk* (Augsburg 1710) and Paul Decker's *Fürstlicher Baumeister* (Augsburg 1711), The decoration in the sacristy at the abbey church of Ottobeuren, particularly the interlacing strapwork, was based in part on these two sources. *Fig. 15* *Fig. 12*

We shall be looking in more detail at the diffusion of styles and instances of direct borrowing in the following chapters – an example is Bagutti's and Artari's hall ceiling at Clandon in England (1730), based directly on an engraving of a representation of Hercules and Iole painted by the Carracci brothers in the Farnese Gallery in Rome. 107–8

Preparatory drawings, models, contracts

When the Ticino stuccoist Francesco Vassalli (fl. 1724–63) was working in England in 1730 he left his work at Towneley Hall, Lancashire, for a time. There were complaints, and in a letter Vassalli indicated that he was glad it was in his power to win back his patron Richard Towneley's favour, 'in that, since I have been in Italy, it is in my power to satisfy you more than I could have done before I returned into Italy . . .'[31] This suggests study of casts, observation of techniques and acquisition of engravings, but also the ability to prepare new drawings for approval.

Generally in the eighteenth century, the architect's control of the design extended to the preparation of these drawings. A number of such drawings for stucco decoration survive, and show that while the architect's ideas of scale and disposition were precise, there was room for the stuccoist to contribute, as for instance in the stance and expression of figures. Two drawings[32] of 1756 by the architect Johann Michael Fischer to show the intended stucco decorations at the abbey church of Ottobeuren suggest the differences in style in the work of the stuccoists. The drawing for the work of Jakob Rauch and Franz Xaver Feichtmayr indicates a more restrained treatment than that for the work of Johann Baptist Zimmermann. The church was drawn in longitudinal sections, and these preliminary plans provide an excellent idea of what would materialize. The drawings were supplemented by the many oil-sketches prepared by frescoists,[33] and by three-dimensional models. *Figs. 1, 10, 13*

When drawings and a model had been approved, it was time to draw up a contract. This safeguarded the patron to a greater degree than the craftsmen. It established the rates of payment (as in the 1724 agreement for the octagon of the Benedictine abbey of Einsiedeln, near Zurich, which provided 2,200 gulden for stucco-work by Egid Quirin Asam and two assistants[34]), the use and cost of old and new materials, and the timetable for the work's progress.

Stucco decoration was one of a sequence of contributions by groups of craftsmen working closely together, moving in at the times their skills were needed.

The decorative team

Enough evidence is available in contracts to allow us to know the normal pattern of team-involvement in at least the South German Baroque churches and their associated buildings. After the submission of plans by the architect and their

Fig. 14. François Cuvilliés (1695–1768), Rococo ornament from *Livres de Portion de Plafonds et d'un Poëlle*, 1738.

Fig. 15. Decorative bandwork from Johann Jakob Biller's *Neues Zierathen Buch von Schlingen und Bändelwerk*, Augsburg 1710.

approval, and the selection of site and its preparation, the masons carried the church to its completion as an undecorated shell. While in several cases this involved the erection of stone or brick vaults (as at Ottobeuren, Weingarten and Vierzehnheiligen), some buildings have a wooden ceiling. The architect's choice of stone, brick or wood vaulting was an important factor for the stuccoist to consider. If a church (or house) were in the more durable materials, fixing-points needed to be established for work in high relief. A timber roof required 'ceiling' over with narrow wooden laths which were nailed to the joists or timber supports before being covered with successive layers of stucco.

XV

Where stuccoists and frescoists were involved, liaison between them needed to be very close to achieve the desired appearance in an interior. Certain ceilings at Augustusburg Castle, Brühl, such as those in the Audience Chamber and the Cabinet, might appear to have been painted first, and then had Guiseppe Artari's gilded stucco of about 1750 set over them like a lavish Rococo frame. However, the reverse is the case, with the painter Joseph Billieux, working in the mid-1750s, carrying his paint carefully beneath the undercut stucco to create this effect. Reflection will indicate that the scratching of the surface and the insertion of nails to hold the stucco-framing was best done on a surface which had not been frescoed. In the choir of the abbey church of Weingarten, painted scrolling breaks out from the pendentives into stuccoed scrolling, and in the Stiftkirche, St Katharina at Wolfegg, the saint carrying a cross, the work of the Wessobrunn stuccoist Johann Schütz, sinks from high relief down into the fresco on the coved ceiling by Franz Josef Spiegler. Such close collaboration might be expected in the work of stuccoist-and-frescoist brothers, like the Asams and the Zimmermanns, but enough examples of it survive to suggest that generally the stuccoist did not feel himself threatened by the painter. The plasterer Edward Goudge in England, however, wrote in 1702 to Sir Thomas Coke: 'I suppose Sir, I need not tell you that for some years past, for want of money occasioned by the War and by the use of ceiling painting, the employment which hath been my chiefest pretence hath been always dwindling away, till now its just come to nothing . . .'.[35]

Fortunately, until the decline of the art of stucco decoration, there seemed to be work for all the artists, and where stucco took hold, we have no evidence that wood-carvers any more than frescoists took commissions away from stuccoists.

The collaboration of architects and migrant craftsmen diffused stylistic influences widely. In decorating the Residenz at Würzburg in the 1740s and 1750s, Balthasar Neumann was able to use the services of a team of talented German, French, Italian and Dutch artists who worked together as a creative group. The stuccoist Antonio Bossi was joined by the sculptor Johann Wolfgang van der Auvera, the woodcarver and cabinetmaker Ferdinand Hund, the Viennese ironworker Johann Georg Oegg, as well as a Viennese potter and porcelain-maker Dominikus Eder, and four painters. This diverse group of workers created the richest and most joyous expression of the Rococo style in a secular setting, while paying close attention to a complex iconographic programme culminating in Tiepolo's ceiling fresco *Homage to the Prince-Bishop as Patron of the Arts under the Protection of the Sun-God Apollo*.

The isolation of a single talent in a craft so essentially anonymous, so subject to influences as stucco, would be a rash exercise.

II

CLASSICAL PATTERNS

Moulded in Antiquity

The earliest recorded use of stucco is in ancient Egypt. In the XVIIIth Dynasty (1567–1320 BC) the sculptors of Tell el Amarna fashioned portraits of Amenhotep III and his son in plaster, together with representations of members of their court.[1] They also used plaster to cover the soft limestone of the celebrated portrait-head of Queen Nefertiti and others. Such heads and statues were given a calm and immortal look, assisted by eyes of crystal, and coloured with delicate pigments. In the New Kingdom stucco played a small but important part as the modelled surface of funerary masks of stuccoed and painted cloth. The back panel of the throne of Tut-ankh-Amun (Cairo Museum) was of wood overlaid with a form of stucco – gilded and modelled gesso – inlaid, and incorporating figures of the king and his wife.

Compositions in low relief were most suited to execution in stucco. These were in use from the peak of the Cretan or Minoan civilization, and particularly at the Palace of Knossos.[2] During Arthur Evans' excavations of the site in 1893 many different representations of the bull were found, executed in painted stucco.

Heinrich Schliemann excavating at Mycenae in 1874 found five royal graves, and among the isolated discoveries, a female head made of stucco. However, the use of stucco in classical Greece was not isolated: as in Egypt, it was used as a covering for limestone to render it smooth for painting and less porous. There must have been whole statues composed of stucco, as the writer Pausanias, in the second century AD, wondered whether the figures of birds on the Temple of Artemis at Stymphalos were in stucco or wood.[3] He also mentions a statue of Bacchus in 'coloured stucco'.

Etruscan civilization from about 700 BC provided the dead with tombs furnished and decorated with frescoes and stuccoes. In the funerary chamber of the Tomb of the Painted Stuccoes in the Necropolis at Caere (Cerveteri) the walls and several square columns have coloured reliefs depicting the possessions of the owner of the tomb or his wife, and portraying real and mythological creatures. These reliefs, coloured in red, blue, yellow, black and white, mark an important step in the use of stucco as a decorative medium. The François Tomb at Vulci is lined with a stucco-facing that simulates marble.[4]

By the end of the second century BC, however, the centre of the ancient world lay at Rome. The 'Masonry style' of fourth-century Greece was the antecedant of Roman stucco of the second century BC, which took the form of reliefs and ceiling-coffering.

Eight schemes of first-century 'coffer-style' stuccoes have been examined in detail. These are in the central Baths at Cales (*c.* 90–70 BC); the House of the Griffins in Rome of the same date, the Baths of the Bull at Civitavecchia, the so-called 'Villa of Galba' near Frascati, the House of the Criptoportico at Pompeii (third quarter of the first century), a late-first-century house on the Palatine, Rome, the stuccoed vault of a tomb at Auximum, and fragments in a Roman house at Spoleto.[5]

The House of the Griffins contains in one of the lunettes a pair of stucco griffins in high relief – some 2 to $2\frac{1}{3}$ inches (5–6 cm) in parts – which give the house its name. Other rooms in the house contain remnants of stucco animals – peacocks and pigeons – and above, simple stucco coffering arranged in squares within broad frames. In the lunettes the white reliefs are set against red backgrounds.

The House of the Criptoportico at Pompeii has figures in action – a pair of cupids wrestle to hold a large quiver, and two more hold torches – but in some cases, no more than a shadow of their outline survives. The coffers here do, however, contain plant-motifs, whereas those at the House of the Griffins are plain. In the *tepidarium* the figure-reliefs, winged sea-horses and dogs and the plant-scrolling are largely intact, and present a rich and varied scene to the onlooker.

One of the most interesting survivals of Roman stucco is contained underground – in the basilica near the Porta Maggiore, dating from the reign of Claudius (10 BC–AD 54). This is built in the form of an apsidal chamber with three aisles, and may well have been used for secret meetings of a sect whose ideals were fashioned on neo-Pythagorean concepts of eternal happiness after death. The stuccoes on the barrel-vaulted ceilings are modelled in spirited and ambitious style with figures, draperies and trees, each layer building up thickly on that underneath. 'Sappho's leap' illustrates the legend of the Greek lyric poetess who threw herself into the sea in the despair of unrequited love for Phaon. The exact symbolic meaning of the stuccoes remains in doubt.[6]

The modelling techniques in the basilica are superior to those in the earlier House of the Griffins, being characterized by firm incised lines. Both, however, are eclipsed by the stuccoes taken from the house near the Renaissance palace of the Farnesina (dating from 30–25 BC, and in the Museo Nazionale delle Terme). The scenes there depicted are not entirely in stucco – there are murals of *trompe-l'oeil* architecture painted in pale pinks, yellow and blues. The low-relief stuccoes show three separate buildings with figures interspersed between them. Two of the buildings have projecting porticoes and are joined by a bridge. We observe that the artist failed to grasp the basic principle of linear perspective, that the receding parallel lines should converge to vanishing point. In each of the buildings two sides are shown, but the parts do not correlate, and the three buildings are not viewed from one point of sight. On one wall a Greek signature *Deleukos epoiei* ('Seleukos made it') appears. Two other rooms were painted black, with a series of stucco festoons, classical figures and landscapes. The decorations offer us an example of upper-class taste at the beginning of the Augustan period.

Stucco was used to decorate the vaults of the Coliseum (begun in AD 70 by Vespasian, continued by Titus and completed by Domitian) with figures of chariots. While little of this decoration now survives, it was drawn in the Renaissance period by Giovanni da Udine.[7]

Similarly, apart from fragmentary remains, our source of knowledge of the abundant modelled stucco decorations and examples of 'grotesque' ornament in the Domus Aurea (Golden House) of Nero of AD 55 is the Renaissance drawings and engravings made by Giovanni da Udine (p. 30). The name of the building derives from the gold leaf that was applied to the stucco mouldings.

Excavations in the 1870s in Rome on the banks of the Tiber and in laying-out the new Quirinal and Esquiline quarters brought several other ornate houses to light – and in Campania the villas at Baia and Cumae were noteworthy. In the Roman Imperial palace at Baia is a series of three rooms, the Stanza di Venere, which were vaulted and stuccoed about the middle of the first century AD. Figures abound, in athletic or restful poses, and although the stucco has deteriorated with long exposure to weather and vandalism,[8] it stands comparison with 'the best Augustan stucco-work of Rome and its environs, as exemplified by the stuccoes from the house near the Villa Farnesina and the decoration of a tomb on the Via Laurentina at Ostia'.[9]

Stucco at its most fluid, and probably 'painted' on with a brush in a thick impasto, won a gradual acceptance during the Augustan period – an example is the decoration of the niche of the Tomb of the Valerii (AD 160) in the Grotte Vaticane under the basilica of St Peter's in Rome. Well-modelled stucco fragments have been recovered from this hallowed site. An example from outside Rome is the flowing decoration of the ceiling of the Stabian Baths at Pompeii (AD 79).

A number of stucco reliefs on funerary monumonuments survive – one (Metropolitan Museum, New York) shows a barbarian kneeling before the Emperor Antoninus Pius (*c.* 138–161).

It was inevitable that decoration with such an ancient lineage, refined and improved by experiment, should spread throughout the wide reaches of the Roman Empire. There are comprehensive evidences of this in the stucco collection of the Landesmuseum at the German 'Roman' city of Trier, and most major excavations throughout Europe have yielded evidence of the use of stucco. In England wall- and ceiling-decorations based on those at Pompeii and Rome have been found at Verulamium, Gorhambury, Bignor and Fishbourne.[10] The stucco was set alongside murals and above mosaic floors.

Stucco was also present in the Mithraeums, which during the first three centuries AD were erected by the worshippers of Mithras as shrines for a religion rivalling Christianity. The most important is that depicting Mithras and the bull under the church of S. Prisca in Rome. The large reliefs and other figures, dating from the latter part of the second century AD, were executed in stucco, but the remains are now sparse.

Stucco of the third century is not plentiful. Various mausolea were stucco-decorated, but there is only one remarkable manifestation – Tomb N in the necropolis of the Isola Sacra at Ostia. A technical improvement here was that the stucco figures were attached to the walls of the tomb by metal armatures, rather than being modelled in low-relief from the background. Henceforward, however, very little was done throughout the Empire, and for more than a thousand years all relics of classical ornamental stucco were buried and forgotten under the ruins of the buildings it had adorned.

The breaking apart of the Roman Empire and the transfer of its power to Byzantium was accompanied by a stylistic trend towards linear forms and the simple frontal delineation of figures. There was continuity of decorative themes such as the scrolling vine. This, motif, indeed, was improved and used for large surfaces.

An important example of pre-Byzantine polychromed stucco of the mid-fifth century is in the Baptistery of the Orthodox at Ravenna.[11] The ornamental arcading in two stages is divided further in its upper reaches into three. The central arches frame windows, and the two smaller ones flanking them contain aedicules with alternating triangular and semicircular pediments. In each is the figure of a prophet set in high relief against an illusionistic background. The episodes depicted in the spaces above the pediments and beneath the arches are based on episodes from the Old and New Testaments. The stucco decoration also includes vine-motifs, rosettes and stylized birds. Further displays of stucco were unfortunately destroyed in the nineteenth century, and are now reproduced in modern frescoes. Panels with stucco animals and human figures were also to be found at Ravenna in the Basilica Ursiana and the Church of S. Croce, but these have largely disappeared. Those at S. Croce were gilded and silvered.

7

There are numerous examples of the skilled fashioning of stucco in Italy during the medieval period. In Brescia in northern Italy at the Church of S. Salvatore (mid-eighth century) the imitation of Byzantine motifs such as the trailing vine, palmettes and rosettes reached a high point of openwork technique.[12] Rome itself, towards the middle of the seventh century, had become a near-Byzantine town, with a number of basilica-plan churches. The stuccoes within them, however, were usually of later date: Byzantium exerted its artistic influence throughout Italy in painting, mosaics, and to a lesser extent sculpture. It was this preoccupation with the sculptured figure that allowed the stuccoists some employment in such churches as S. Maria in Valle in Cividale del Friuli. The seventh-century stucco figures of the saints in jewel-studded robes and the vine-leaves and *rinceaux* make a powerful, flowing statement. In Milan at S. Ambrogio there are similar stucco decorations – modelled with vigour and well-incised lines – which find parallels in tenth-century work elsewhere in Europe, as at Reichenau on the shores of Lake Constance.

In Germany stucco is important in the medieval period, as it was to become in the fervent years of the Baroque. The abbey church at Quedlinburg has stucco tomb-slabs dated to about 1130 that provide a place for decoration, but these are overshadowed by the fine, attenuated figures of the Apostles and saints at Gandersheim (*c.* 1120), Hildesheim (*c.* 1186, some at the East Berlin Museum) and Halberstadt (end of the twelfth century), all magnificent examples of Gothic art. The reliefs at Hildesheim, which depict the saints in carefully folded robes within semi-circular-headed canopies divided from each other by pilasters,[13] are precise and formal. Hildesheim owed its prosperity to the fact that in 822 it was made the seat of the bishopric which Charlemagne had founded a few years before. The art of working in metals was greatly encouraged there, and stucco was useful for the first stage in modelling.

8

The patterns of Islamic art introduced to Spain during the Middle Ages were based on geometry and on the formalized depiction of natural forms such as the five-lobed vine-leaf. From an abstraction of the shapes made by plant-forms, the arabesque was derived. This flowing motif – often executed in low-relief stucco – allowed the creation of patterns in which there is no beginning and no end. One of the most interesting of these patterns was interlace, in which the linking ribbons cross each other in a three-dimensional arrangement. Stucco was used extensively, also, for the calligraphy that is such a universal and important element in the adornment of Islamic architecture. Islamic craftsmen excelled at linking straight lines or curves, and used this ability to create stucco stalactite-domes in their architecture – for example at the Alhambra, Granada.

Stucco is one of the materials most frequently employed for decorative purposes in the Great Mosque of Cordoba, at the Alhambra at Granada, the pavilions of the Generalife Gardens (north-east of the Alhambra group), and in the Aljaferi Palace in Saragossa. The interior of the mosque at Saragossa was decorated in stucco to the exclusion of almost any other material, but in the citadel and palace of the Alhambra, as well as in the Alcazar at Seville, patterns were created in both stucco and tiles.

In the dry climates of the Mediterranean lands stucco could be used externally, as for window grilles. The play of light – the reflections, the shadows – is an intrinsic element of Islamic architecture. In strong sunlight, pierced stucco appears almost translucent; as the sun moves across, it casts its shadow-patterns on to surfaces already richly patterned.

Renaissance Ornament

In 1488 in Florence, Cardinal Giovanni de' Medici was seeking for sculpture of classical origin. Giovanni, created Cardinal at the age of thirteen, had acted upon advice from his father Lorenzo Il Magnifico in collecting objects of antiquity and rare books rather than silken clothes and jewels. The Cardinal caused excavations to be made in Rome over several years among the many ruins of the classical city. In particular he concentrated on the 'Domus Aurea' – the 'Golden House' of Nero[1] – known for its great size and costly decoration by the descriptions of the Latin writers, and particularly that of Suetonius in his *De Vita Caesarum*. The result of the excavations made there on the Cardinal's behalf at the turn of the century was not only the discovery of statuary, but also the more surprising reappearance of rooms or 'grottoes' which were painted, decorated with mosaic and stuccoed.

The 'Volta Dorata' in the centre of the Domus Aurea façade was decorated with stucco relief-figures of mythological groups, framed by square and rectangular panels. The façade was picked out in gold. Villas had become in the last centuries BC the focus of Hellenistic luxury. In Nero's villa, the favourite design in all the painted and stuccoed decoration was of arabesques, sphinxes, griffins, vases and candelabra.

The Cardinal allowed artists to be constant visitors to the decorations of the western range as it emerged from beneath the terrace of the Thermae of Trajan. (Apart from some exploration in the 1770s, the eastern range was not uncovered until 1912.) Two artists who were allowed, about 1500, to view what the Cardinal's men had found were the painter Raphael (1483–1520) and his assistant Giovanni da Udine (1487–1564).

Raphael realized that imitation of the ancient *stucco-duro* would provide a superb new form of interior decoration. Rome was ready to be transformed under the patronage of the new Pope, Julius II, who had been elevated in 1503. He had asked the architect Donato Bramante to start building the Belvedere Court of the Vatican, and to rebuild the dilapidated church of St Peter as a great basilica worthy to house his own tomb. Julius died, however, in 1513 with many of his schemes incomplete, and Raphael's patron Giovanni de' Medici, intent on succeeding as the next Pope, saw to it that Raphael and Giovanni da Udine, among the most talented of their day, could pursue their investigation under his protection.

First it was necessary for them to examine more caves or 'grottoes' in the great complex of the Domus Aurea – estimated, with its grounds, to cover over two hundred acres.[2] According to the report of Giorgio Vasari, chronicler of artist's lives, both were astonished by the grotesque paintings and scenes and figure-reliefs in stucco which they found. As the decorations had lain buried in the earth, immune from the destructive exposure to the atmosphere, it was small wonder that they amazed the artists, as Vasari noted, by their 'beauty, freshness and excellence'. Vasari confirms that Giovanni da Udine, after drawing the ornament 'over and over again', then puzzled out alone the secrets of the ancient stucco-composition. He succeeded at a propitious time, for Raphael's patron was elevated as Pope Leo X in 1513. The Pope's immediate task was to authorize the continuation of the construction of St Peter's and the loggias of the papal palace which had been started by Bramante and others under the patronage of his predecessor, Julius II. At Bramante's death in

1514 the way was clear for Raphael and his team to engage themselves on these decorations, chiefly fresco but including many in white stucco. Vasari's description of what was done under Giovanni da Udine's direct supervision in the loggias is concise. Raphael 'employed Giovanni to cover the vaulting with stucco, with rich ornaments surrounded by grotesques similar to those of the ancients, full of the charming inventions and the most varied and curious things imaginable. He did the entire decoration in bas- and half-relief, introducing therein scenes, landscapes, foliage, and various friezes, endeavouring to achieve the utmost of which that art is capable . . . his numerous apprentices learned under him, and then scattered through the provinces. . . .'

This style of bright painting intermingled with relief-stuccoes, copied from the ancient stuccoes discovered in the excavations, was to have a profound effect on later decorations in Rome, including those by Raphael's team for the Pope's cousin Giulio de' Medici, and, as we shall note, at the Villa Lante and the Casino of Pius IV.

The Pope's cousin Cardinal Giulio de' Medici had decided, probably sometime around 1515, to build a secluded residence near Rome on the green slopes of Monte Mario, using the services of Raphael and Giovanni da Udine. This palace, now known as the Villa Madama, contains some of the finest early examples of the revived art of stucco decoration. Who was ultimately responsible for the design of the building is a complex question into which we need not enter. Raphael and his talented pupil Giulio Romano were probably concerned,[3] although Raphael died early, at the age of thirty-seven, in 1520. The stucco decoration was almost certainly supervised by Giovanni da Udine, assisted by Giovanni Francesco Penni and Baldassare Peruzzi. The work is exquisite, and, based directly on that in the Golden House, conformed with Raphael's abiding intention to recreate the atmosphere of a classical Roman interior. The garden loggia has its domed vaults covered with stucco and grotesque painting.

v, 9

As the Cardinal was in Florence for much of his time, the decorations[4] were supervised by his friend Bishop Maffei, who suggested to the artists that the stories to be depicted should be well known and pleasing to the Cardinal. The fables of Ovid took precedence over any religious themes. The dome of the central bay in the loggia contains stucco reliefs of the 'Four Seasons' in the forms of Proserpine, Ceres, Bacchus and Vulcan. The west cross-vault contains a stucco figure of Neptune, while that to the east contains a corresponding figure of Amphithrite. In the half dome of the exedra a series of ten stucco reliefs depict incidents from the story of the love of Polyphemus for Galatea. In white, with some gilding, these are surrounded by a connecting framework of low-relief stucco ornamentation of vases, griffins and small *putti* amid acanthus-scrolling. The domes are surrounded by the brightest grotesque paintings in dominant blues, reds, greens, fawns and yellows. What Raphael and Giovanni da Udine had seen in the ancient 'grottoes' had been effectively adapted for use in the grandest of Renaissance villas. The Cardinal, who was elevated as Pope Clement VII in November 1523, saw to it that his arms, both as Cardinal and Pope, were often incorporated in the decorations. Giovanni da Udine signed his own name on one of the pilasters in the garden corridor connecting the loggia and the circular court, alongside his patron's name and the date '1525'.

The Pope's favourite recreation was to visit his villa as often as possible, to supervise work on the gardens and to enjoy the quiet retreat which was the reason for the erection of most villas. Those in Rome served, too, as evocations of the classical or mythological past.[5]

Giulio Romano

It was natural that Giulio Romano (1499?–1546), trained under Raphael and with experience of work for the new Pope Clement VII, should become an advocate of the newly rediscovered grotesque form of painting and stucco decoration. He soon had occasion to display the knowledge of its use he had gained while working at the Villa Madama under Raphael's direction, and after the artist's early death, under the experienced Giovanni da Udine.

The Villa Madama decoration was more or less finished by 1525, thus releasing Giulio Romano to move north to Mantua into the service of Duke Federigo Gonzaga. The Duke wanted him to design and construct his *villa suburbana*, a short distance from the Ducal Palace at Mantua. What they achieved together was the Palazzo del Tè, set on the Isola del Tè beyond the city walls. Mannerist in style, the palace was 'the greatest commission of Giulio Romano's entire career'.[6]

The term *maniera* (from which we derive 'Mannerism') has been used in Italy from the late fourteenth century to mean 'individual style'. It may in over-simplified terms be described as a deliberate breaking of the 'rules' of classical art established at the High Renaissance. The style was current from about 1515 to 1600.[7] It is characterized by the use of motifs out of their original context, and by subjective and emotional displays of complex forms and perspective in architecture, and attenuated figures and strident colours in painting. In stucco Mannerism is less easy to define, but figures tend to show the same attenuation as in painting, and to assume twisted and convoluted poses. Giulio Romano especially, Vasari and even Palladio were exponents of the style.

The Palazzo del Tè was complete by the middle of the 1530s, and used by the family for its leisure. The seventeenth century saw its decline following the sack of Mantua in 1623, and further despoilation in the early eighteenth century hastened its decay. It was not until the late 1770s that efforts were made to care for the remains of the building.

The Palazzo del Tè is square in plan, with four inward-facing façades around a rectangular courtyard, the east side looking over walled gardens containing fountains decorated with stucco. The bays of the façade have heavily rusticated door- and window-surrounds and column-bases, in stucco simulating stone.

The stuccoed rooms are for the most part in the range on the eastern and northern sides of the courtyard, with the most flamboyant, the Sala degli Stucchi, in the south-east corner.

While some effort of imagination and study of the relevant documentation[8] allows a partial reconstruction of what was originally intended, it is the brighter colourings favoured by the eighteenth-century repainters and stucco-restorers that we now see. The style ranges in mood from the joyful, bright interiors of the Sala di Psiche to the restrained elegance of the Sala degli Stucchi.

The team of stuccoists included the painter Primaticcio, Andrea and Biagio de Conti, Niccolò da Milano and Giovanni Battista Mantovano. Regrettably no documentation survives for the Sala degli Stucchi. It was decorated some time between 1529 and 1531 – the two main authorities differ as to whether the room was ready for the visit of Charles V to Mantua in April 1530, and on whether Primaticcio – who was at this point to be an important carrier of the Mannerist style to France – was involved in its stucco decoration. The ceiling is divided into twenty-five square coffers containing bas-relief stucco figures portraying religious, historical and mythological subject matter. 'The Forge of Vulcan', 'Mercury and Venus' and 'Proserpine and Dionysus' were set against coloured grounds (which are now light green) and like the other subjects, are mounted on the surface of a barrel-vault ceiling.

Colour plates, pages 33-36

I Loggia of the Villa Madama, Rome, 1519–23. The decoration, in paint and stucco, is by Giulio Romano and Giovanni da Udine and reflects the influence of the Golden House of Nero, then recently discovered. This view, looking west, shows the three main vaults, the central one higher and surmounted by a dome. Here are concentrated a host of figures drawn from classical mythology and from heraldry: Neptune, Jupiter, Juno, Pluto and Proserpine, with the four seasons circling the Medici arms. The villa was built for Cardinal Giulio de' Medici, later Pope Clement VII, who achieved his ambition of creating a setting with all the 'leisure, comfort and beauty . . . that one could wish'.

II Scala d'Oro (Golden Staircase) of the Ducal Palace, Venice. The staircase, built after 1559 to the designs of Alessandro Vittoria, connects the loggia storey to the Sala del Collegio. Vittoria was essentially a Mannerist, but the heavy white and gilded stuccoes and low-relief figures in the tunnel vault show him to be still an upholder of Renaissance forms.

III Detail of a column from the courtyard of the Palazzo Vecchio, Florence. The gilt stucco ornamentation with its mask-faces, heavy swags and nonchalant figures was added in 1565 under the direction of Vasari. It honoured the wedding of Cosimo de' Medici's son, Francesco, to Joanna of Austria. The combination of classical gods, winged herms, satyr masks, foliage, cornucopiae and garlands makes up a language that goes back, via Raphael, to ancient Rome, and leads forward to Baroque and Rococo. Cosimo, in a letter of October 1565, warned his son, now prince-regent, that artists were difficult to handle and 'always take longer than they say they will'. But the urgency of the wedding-preparations, and Vasari's careful supervision, saw at least to the completion of much of lasting significance.

IV Detail from the Chambre de la Duchesse d'Etampes at Fontainebleau, *c.* 1541–4. The Mannerist style evolved here by Rosso and Primaticcio—the so-called 'School of Fontainebleau'—was to determine the course of French art for the rest of the sixteenth century. These elongated figures, self-conscious poses and wealth of decorative detail are typical of Primaticcio's work. He had learned his skills in Mantua under Giulio Romano, and took the secrets of stucco manufacture to the French court only ten years or so after they had been rediscovered by Raphael and Giovanni da Udine.

II

III

figures and heavy swags, bear a slight stylistic relation to those executed in 1726 in England by the Ticino stuccoist Giuseppe Artari at Houghton Hall, Norfolk.

Until the liberation from the Turks, who occupied Hungary for most of the seventeenth century, opportunities for architectural development there were limited. The most impressive stucco is that by Andrea Bertinelli in the Festsaal of the Burg at Sárvár (1653), where heavy ribs and pendentives surround battle-frescoes. The ceiling of the Jesuitenkirche at Gyor (1635–42) has stucco of a derivative form, with linked scrolls placed diagonally from each corner of the ceiling to a central frescoed panel. At Eisenstadt the medieval castle was replaced to designs by Carlo Martino Carlone. It was therefore not surprising that external stucco decoration, known in Italy and in his native Ticino, should be used for the courtyard façade, *c.* 1670.

Through decoration in Yugoslavia was mostly in wood, an unknown stuccoist formed, in Rococo style, the attractive pulpit and side-altar in the church of St Rochus at Šmarje, near Jelše.[58]

Sweden

It was a German stuccoist, Daniel Anckermann of Mecklenburg, who, about 1630, introduced the art of stucco-working in Sweden. Coming on from Riga, he worked on the stucco battle-map in the funerary chapel to Admiral Karl Karlsson Gyllenhielm in the cathedral at Strängnäs (1649–52). He also seems to have been responsible for the remarkable but stilted conception in the late 1640s of the stucco equestrian figure of Herman Wrangel at Skokloster. He had left Sweden by 1656.[59]

In the castle at Skokloster three ceilings were stuccoed in 1663–4 with extremely heavy geometric ribs, and reliefs featuring St George and the Dragon. These are by an Italian stuccoist, Giovanni Anthoni, working in collaboration with a German, Hans Zauch. They also trained one of the few Swedish stuccoists, Nils Eriksson. Giovanni and his brother moved on in 1668 to decorate the castle of Djursholm, near Stockholm, but Giovanni died, and its completion was left to his younger brother. Some of their work was based on engravings by Jean Lepautre which showed the formation of acanthus and laurel-wreath scrolls. *Fig. 9*

Two Comasques to journey to Sweden were Giovanni and Carlo Carove. In April 1664 Giovanni signed a contract to provide stuccoes in the royal palace of Drottningholm. These are formed of heavy cartouches and shells, and formalized coats-of-arms over doorways. Giovanni Carove left Sweden for a time in 1667, but Carlo Carove in that year decorated the funerary chapel in the church at Floda, a hundred kilometres west of Stockholm, with an impressive array of figures. When Giovanni returned in the autumn of 1667, he and Carlo both worked at Drottningholm, but by 1673 Giovanni had returned to Germany.[60] Carlo stayed in Sweden until his death in 1697, and also trained the Swedish stuccoist Simon Naucleurs, who worked with him at Mälsaker (1676) and Tyresö (early 1680s).

When Carlo Carove died in 1697 the stucco decoration of Drottningholm was continued, under Nicodemus Tessin the younger, by his compatriot Giuseppe Marchi, who had started at the royal palace in 1694. He paid close attention to motifs found in Cesare Ripa's important sourcebook, *Iconologia*, which had been published throughout Europe in succeeding editions since its first appearance in 1593. Marchi worked in a similar style to Pietro Peretti and Giovanni Galli. All three were adept at modelling fine acanthus-scrolling, and there is little to choose between the quality of the Peretti-Galli work in SS. Peter and Paul at Vilna, and that by Marchi under windows and balustrading at Drottningholm (1701–2).

Marchi's activities elsewhere included the depiction of birds, shells and other naturalistic motifs at the palace of the statesman Johan Gabriel Stenbock (1704). These are modelled with great fluency – the stucco almost appears to 'flow', and there is less concern with keeping the ornament within the area defined by the ribs. Marchi's work in the following year at Stockholm University is as crowded – Charles XII is personified as Mercury, part of an elaborate allegory of commerce and navigation – but the ceiling has finer shells and acanthus-scrolling to commend it.

French influence had reached Sweden with Nicodemus Tessin at the royal palace in Stockholm in the 1690s, when in a role likened to that of Le Brun in France, he turned for inspiration to Versailles. He also urged the Swedish envoy in Paris, Daniel Cronstrom, to engage three or four French craftsmen for service in Sweden. One of these was the sculptor Bernard Chauveau, a pupil of Girardon, who arrived in Stockholm in 1693 with the Italian stuccoist Pietro Pagani. Some of their decoration was destroyed in the fire of 1697, which consumed the palace and destroyed its chapel. That in the Gallery, the Salon de Psyche and the audience chamber survives, is finely gilded, and set out on the ceiling and over the cornice with a bravura worthy of its Baroque setting. In the Gallery Chauveau and Pagani are 65 credited with the execution of twelve figures, coats-of-arms, cartouches, royal insignia and bas-reliefs, for which they were recompensed handsomely. Changes in taste occasioned some later replacement of the decoration by paintings in a Berainesque style, but Sweden's involvement in the wars of the late seventeenth and early eighteenth centuries sealed the castle from alterations until the introduction in 1727 by Carl Harleman of spirited Rococo motifs.

England

The Jacobean style of the early seventeenth century in England was largely derived from vigorous Elizabethan plasterwork which was often modelled *in situ*. The influence of the Continent became discernible only in a growing use of strapwork and arabesque, cartouche and mythological panel.

Pattern and instruction books made their appearance slowly, but eventually no gentleman could afford to ignore them in selecting the decorative motifs for his plasterer to copy. Books by Abraham de Bruyn (1584), Wendel Dietterlein (1598) and Jan Vredeman de Vries (1563) provided a small, sophisticated repertory of ideas. An increasing use of the great biblical themes and citation of texts followed the 1611 edition of King James's Bible, while allusions to classical knowledge were prompted by the English translation of Serlio's *Architettura*. Modellers of the time also used engravings in herbals and emblem books as sources for ornament.

Two of the more accomplished of the early seventeenth-century plasterers were Richard Dungan and James Lee. Dungan worked at Knole, the Kent house of the Sackvilles, and at the Tudor palace of Whitehall in London, and Lee was employed at Hatfield House in Hertfordshire by Robert Cecil, first Earl of Salisbury, who had secured the accession of James VI to the English throne in 1603. The work at Knole, for the King's Lord High Treasurer, the Earl of Dorset, consists of geometric patterns of interlaced ribs with stylized flowers which had been separately moulded set in the spaces between them. It provides the supreme example of plasterwork of the period, as Lee's work at Hatfield, presumably of a similar style, has now disappeared.

A ceiling which rivals the Knole work is that of about 1606 from the Old Palace of Bromley-by-Bow, now in the Victoria and Albert Museum, London. It is planned in intersecting squares and quatrefoils, and eight pendants (a dominant feature of Scottish interiors of this period) hang from the points of intersection. Circular medallions are surmounted by wreaths and winged cherubs' heads, and full-faced, bearded busts represent three of the Nine Heroes or Worthies. The use of the Bible as a source has nowhere a more lively example than at Lanhydrock, Cornwall, where the twenty-four sections of the Long Gallery ceiling, *c.* 1635, illustrate events of the Old Testament, from the Creation to the burial of Isaac. No record of the exact date of its creation or the plasterer's name has survived. What the anonymous host of plasterers did was to cover such ceilings with precise low-relief work, needing little support other than adherence to the underlying structure of wooden laths.

A rare survival is the sketch-book of one John Abbott of the 1670s (and four of his tools), which shows how a West Country plasterer modified London ideas to suit the Somerset, Devon and Dorset interiors he decorated. The elaborate plaster decoration at Forde Abbey, Dorset, may be by John Abbott and his father. The acanthus is robustly modelled, and stylized mermen divide or terminate the scrolling. The work shows little knowledge of classical precedent, and may be regarded as typical of what was accomplished far from London, or for patrons who had not travelled or acquired the latest books.

66

Although the distinguished architect and designer Inigo Jones gave early advice about the building of Coleshill House, Berkshire (destroyed by fire 1952), its form was decided by a country squire, George Pratt, and his gentleman-architect cousin, Sir Roger Pratt, about 1650–2. The heavily ribbed ceilings, with plaster over a wooden core, carried a profusion of individually moulded fruit and flowers, set above rich cornices – one containing twenty-four plaster shields connected by thick swags. They were probably the work of the elder John Grove, a London plasterer of considerable ability. His date of birth is unknown but was probably about 1610. In 1660 Grove became Master Plasterer to the King's Office of Works, which superintended building and repair of the Royal Palaces.

67

While the influence of Sir Roger Pratt, a much-travelled man, was encouraging a wider concern with classicism, he was quick to point out that it was not necessary 'to proceede to a rash and foolish imitation' of Italian models. In any case, money was becoming scarce because of the Civil War, and regular work for plasterers had to wait for the Restoration of the King and the return of the court from its exile abroad.

An almost unbroken age of grandeur began with the Restoration of Charles II in 1660, when craftsmen threw off the stylistic constraints of the preceding sixty years – none more actively than the artificers of the Office of Works, which included the teams assembled by Grove the Master Plasterer.

Grove's plasterwork at Coleshill now set the standard for fifty years. At his death in 1676, Grove's son, also John, succeeded his father and was to serve the Surveyor-General, Sir Christopher Wren, for a further thirty years until his own death in 1706. Their work was of unvarying quality – profiled ribs set out with unerring accuracy, and with fruit, flowers and simple heraldic devices decked about them in a free naturalistic way.

Royal work went on at Whitehall, Windsor, Winchester, Newmarket and elsewhere – much of it now destroyed – while plasterers who trained and worked in London contributed much. The Worshipful Company of Plaisterers there had fifteenth-century origins, and had received confirming charters from Henry VII and Queen Elizabeth. In the many Livery Halls, erected after the Great Fire of 1666, and in Christopher Wren's City churches, plasterers found excellent new settings to display their art. By 1670, Parliament had decided to rebuild fifty-one of the eighty-seven churches destroyed in the Great Fire, and John Grove junior, along with his partner Henry Doogood, was employed as plasterer at forty-three of them, receiving overall the large sum of £8,060. Without variation their work was in white, and characterized by accomplished scrolling and setting of fruit around geometric displays of profile ribs. There is a handful of lesser names: John Combes, Robert Horton, Thomas Meade (his lovely vestry ceiling of about 1675 at St Lawrence, Jewry, was destroyed in the Second World War), Robert Dyer and John Sherwood, who between them served Wren and the Church vestry officers at another ten City churches.

Henry Doogood, John Grove's partner, had first come to Wren's notice in 1663, when he used him to execute the simple but distinguished ceiling with one great oval rib at Pembroke College Chapel, Cambridge. What plasterers like Doogood did was dictated by the precise rules of the Office of Works. The methods and materials were decreed in every contract – 'first lay a coat of lime, sand and hair, and scratch it over to make a good key for the second coat'. The stuff (or material) was 'to be good, well wrought in the best manner and beaten six times to render it smooth'. The final coat was of plaster containing white kid's hair. The whole was measured finally, after smoothing or 'floating', with a twelve-foot rule.

Doogood and the Grove family almost had the monopoly, but Wren did patronise a few other plasterers. One of them, James Pettifer, joined with Robert Bradbury in 1675–6 to provide some of the finest ceilings outside London, at Sudbury Hall, Derbyshire. The Long Gallery ceiling in particular, over one hundred and twenty-six feet (42 m) in length and twenty-three feet (7 m) broad, is interesting for the skill shown in setting out the several ovals and squares, and also as a rare example of what a country squire (George Vernon) thought was fitting to decorate the interior of his brick-built and somewhat archaic house.

Of recent years we have come to know much more about the plasterer Edward Goudge, whom the architect Captain William Winde called in 1688 'the beste master in England in his profession'. We can assume a long but casual acquaintance with Wren, from the time, about 1680, when Goudge introduced to the Surveyor-General the almost unknown Nicholas Hawksmoor who was to become Wren's (and later Vanbrugh's) assistant and a talented architect in his own right. Goudge's principal private work of the late 1680s still surviving in England is the accomplished ceilings at Belton House, Lincolnshire, and Castle Bromwich Hall, Warwickshire. The fruit and foliage, individually modelled, is heavy and set with overwhelming complexity into the broad flat ribs of his ceilings. Winde said that Goudge made all his own designs – that for his (now destroyed) ceiling at Hampstead Marshall, Berkshire, of 1686 survives in the Bodleian Library, Oxford. It shows the usual dominant central oval, with two rectangular panels, one at each side, with a rich riot of acanthus-scrolling surrounding heraldic shields. They may owe a little to the engravings of foliation by Jean Lepautre in France. By the turn of the eighteenth century Goudge was finished as a plasterer, his profession was in decline, yielding before the colourful palettes deployed to create rich frescoes. *Fig. 9*

These last years of the seventeenth century had allowed, by the development of a more refined taste and an increase in travel, a finer English plasterwork than ever before. The plasterers found difficulty, however, in ridding themselves of stiff, geometric borders, and were hesitant in handling life-size figures. Sir Roger Pratt had instanced in his notebooks how ill-suited to this sort of sculptured work was the talents of English plasterers. None of them knew the accomplishments of Bernini's and Borromini's stuccoists in Rome, or those active at the Theatinerkirche in Munich. It is significant that the portrayal of the figure, dependent on knowledge of engravings and frescoes of Italian origin, is absent in London work of the seventeenth century. Most of the work was based on observation of nature, of flowers, fruit and animals. It was moulded on tables at ground level, and taken to scaffolding-level to fit into position.

The 'heathen Gods, Emblems, Compartiments &c' which John Evelyn had noticed at Nonsuch on 2 January 1666, and pronounced 'must needes have ben the work of some excellent *Italian*...' are found again only with George I, and the *stuccatori* who came from Lugano and Como.

Spain

In the second half of the seventeenth century Spanish Baroque threw off its older stylistic allegiances. In particular the achievements of Jose Benito de Churriguera (1665–1725) heralded a new 'Churrigueresque architecture' which smothered its structures – not always designed by members of the family – in an excess of stucco decoration of Hispano-Moorish origin. This is characterized by considerable complex relief, and recession in interlocking patterns. Jose's fame was established by the altar in the Church of the Incarnation in Madrid (1693), with its twisted columns derived from Bernini's baldacchino in St Peter's in Rome. Jose's sons, Matías, Jerónimo and Nicolás, helped him with his many commissions, and two other members of their family were active at the new cathedral in Salamanca from 1713. The group's interest in light and shadow finds its most effective expression in the crowded stuccoes in the sacristy of the Carthusian monastery (Cartuja) at Granada, designed by Luis de Arévalo.

In Seville the exuberant architectural work of Leonardo da Figueroa (*c.* 1650–1730), with its emphasis on undulating cornices, statues of saints, patterned columns and glazed tiles, allowed little room for stucco. It did, however, encourage the spread of the Baroque style to the Spanish-American possessions. Italian-style stucco applied to the ceilings of churches in Seville and Granada in the early 1650s was copied in Mexico by the stuccoists of the Rosary Chapel in Sto Domingo at Puebla (who also worked on the interior of Sto Domingo at Oaxaca in 1657). Here the work was sometimes more advanced in the creation of interesting patterns than that being done in Europe, and in Spain itself. The ensemble of polychrome tiles, coloured stones, undercut sculpture (which echoed Pre-Columbian themes), was enriched with a local repertory of naturalistic motifs. Stucco is confined to a few sites, and is rarely the sole decorative element – the upper part of the altar of Sta Clara at Queretaro (*c.* 1730), for example, is covered in decoration part-stone, part-tile, part-stucco, whose excesses defy description.

The same concern with complicated and tortuous ornamentation is found on the stucco portal of the palace of the Marquis de Dos Aguas in Valencia, from the 1740s. A fine work, but with all the excesses characteristic of Spanish Baroque, it was designed by the painter Hipólito Rovira Brocandel (1693–1765), and executed *c.* 1740 by Ignacio Vergara (1715–76). His father Francisco (1681–1753) and he were much influenced by the German architect Konrad Rudolf, a pupil of Bernini, who had begun the Baroque façade of Valencia Cathedral in 1703. 69

The principal stucco ceilings in Portugal are in the Palacio de Correio-Mor at Loures, near Lisbon. The palace was developed from an ancient farmhouse by the Postmaster-General (Correio-Mor), José Antonio Matta, in 1735, a time when the influence of John V was strong. All four rooms on the first floor have elaborate stucco ceilings by a Comasque stuccoist, Carlo Rossi.

While Spain and Portugal profited like the rest of Europe from the willingness of artists and craftsmen to travel, even gilded stucco, versatile for creating rich decoration, could not compete for long with wood inlaid with other woods imported from Spanish and Portugese possessions in the Americas. There was no transfer of influence in respect of stucco to elsewhere in Europe, and neither country made any use of the French engravings by Berain and his followers which were to be so influential in the forming of stucco-patterns in Germany.

We have referred above (p. 60) to the diffusion of a lighter style of decoration, under the influence of engravings and craftsmen from France, by the end of the seventeenth century. In the early years of the eighteenth century the *genre pittoresque* was started in France through the work of a number of artists born around 1690, such as Juste-Aurèle Meissonier, Nicolas Pineau and François Cuvilliés. The taste for asymmetry became the *raison d'être* of their art, and their books of engravings, issued in the 1730s, were to be important influences on the decoration of Rococo interiors created for the Electoral court at Munich. The interlaced ribbons (*Bändelwerk*) and grid motif (*Gitterwerk*) became, in Austria and Germany as elsewhere, a sign that not only had the new mood arrived from France, but the Lombard stuccoists who understood it best and used it to such advantage were present also. *Fig. 14*

The Roman precedent

Stucco reached a high level of accomplishment in the Roman Empire, but comparatively little has survived.

1,3 One of the most interesting survivals of early first-century stucco decoration in Rome is in the subterranean Basilica di Porta Maggiore (left). One low relief (below) illustrates the legend of the Greek lyric poetess Sappho in spirited and ambitious style.

2 The remains of white stucco lunette decoration, 90–70 BC, of a pair of griffins from a house on the Palatine in Rome.

Art of the 'grottoes'

Ancient remains discovered during the Renaissance were beneath ground level and were therefore at first thought to have been subterranean rooms: hence the term *grotteschi*, derived from *grotto*.

4–5 The stucco landscapes with figures from the house near the Farnesina, Rome (now in the Museo Nazionale delle Terme), dating from early in the Augustan period, 30–25 BC, show the modeller's attempts at perspective, relationships of scale and depiction of figures. The illustrations also give some indication of the comparatively thin coats of stucco (10–15 mm) forming the background, and the building-up of the relief figures which were incised, *in situ*, with modelling tools

6 Vault decoration of the Stabian Baths at Pompeii. Stucco at its most fluid was probably painted on to the surface as a thick impasto. This first-century example seems to have been formed by a combination of brush-painting, and, for the roundels and octagons, the scribing of their shapes with a profile mould, and indenting their floral and egg-and-tongue enrichments with a press-mould.

Decline and revival

Use of stucco lingered on until the early Middle Ages as a cheap substitute for stone, but without fresh technical or artistic inspiration. When Roman stucco-work was rediscovered in the Renaissance it initiated a whole new era.

7 The interior of the pre-Byzantine Baptistery of the Orthodox at Ravenna, built *c.* 450, is richly decorated in polychromed stucco, marble and mosaic. The stucco-work is in the upper of the two superimposed arcades. The central arch contains a window, the two flanking arches aediculas with alternate triangular and semi-circular pediments. In each aedicula is the stucco representation of a prophet. Above the pediments, stylized birds and animals show the ability of the stucco-modeller.

8 Apostle from the choir screen of St Michael's, Hildesheim, Germany, late twelfth century. Within the Romanesque church are two compositions on the choir screen. Set into blind arches they incorporate representations of the Virgin and Apostles standing under canopies, with cupolas and buildings in perspective between them. Other stucco images of saints, *c.* 1186, have been removed and are now in the Staatliche Museum in East Berlin.

9 Loggia of the Villa Madama, Rome, 1519–23. The white stuccoes set amid coloured painting (*see* pl. I) are disposed over the soffit of arches, pendentives, domes and walls with rare skill. These decorations had a far-reaching influence on the stucco of other countries during the succeeding years.

The school of Wessobrunn

During the seventeenth century, the Benedictine abbey of Wessobrunn became one of the two leading centres for training stuccoists.

54 A corridor in Wessobrunn Abbey, typical of the Wessobrunner style: a lavish use of acanthus foliage, great shells, stylized flowers and cherub-heads.

55 Detail from the ceiling of the abbey church of Obermarchtal, 1689–92, by Johann Schmuzer, one of the most outstanding Wessobrunners. Foliated ovals and roundels are connected and divided by simple foliated ribs.

56 Priory church of Friedrichshafen, by Johann Schmuzer assisted by his two sons Franz and Joseph (restored 1950). The stuccoes are richer than those at Obermarchtal, with the cross-vault in each bay containing a great shell. Cherubs' heads, and shallow oval and rectangular cartouches are set into a more restrained acanthus framework.

Figures and flowers

As techniques became more and more assured, stuccoists crowded together motifs from a variety of sources, producing dense masses of ornament that often overflow their frames.

57–8 Pilgrimage church of Vilgertshofen, 1687–92, by Johann Schmuzer. The general appearance of the church may be noted in pl. VI, but these two views allow a closer look at the complexity of the ceiling ornament. Around a balanced arrangement of frescoes run flat decorated ribs carrying acanthus and winged cherubs' heads. Many of the acanthus fronds may have been moulded and then set into position while still damp. This would allow the leaves to be twisted into the most appropriate position, after which the veining would be incised with a pointed modelling-tool. Finally the whole ceiling would be painted with 'whiting'—a cream-like solution of lime in suspension.

59 Schönenberg pilgrimage church: the ceiling of the vestibule, *c.* 1682, possibly by the Schmuzers. Here the whole repertoire of stucco is brought into play: cherubs, standing on acanthus, support wreaths of leaves and flowers, other cherubs hold shells, while delicate arabesque ornament weaves through the spaces, harmonizing and uniting.

60–1 These two allegorical figures of America (left) and Africa (right), *c.* 1685, are on the Kaisersaal ceiling of Schloss Alteglofsheim, near Regensburg. 'America' wears a feather head-dress and carries a bow and arrow, 'Africa', mindful of unremitting heat, a sun-umbrella. These delightful figures (two of four, with 'Europe' and 'Asia' completing the quartet) were based on engravings in Ripa's *Iconologia*. They form part of a Wessobrunn-style ceiling, with elaborate acanthus scrolling, although it was probably executed by one of the Comasques working with Giovanni Battista Carlone at Passau Cathedral (*see* pl. 47).

Bohemia and the east

Bohemia and to a lesser extent Poland were experiencing the same upsurge of building as Bavaria and Austria, and relied equally on Italian expertise.

62 Ceiling of the Castle chapel of Nachod, near Prague, *c.* 1654. The castle was reconstructed in 1653 by Carlo Lurago, assisted by two stuccoists from Como.

63 The Sala Terrena of the Archbishop's Residenz at Kromeriz. The stuccoes, *c.* 1690, by Baldassare Fontana, decorate niches containing life-size statues of Diana and other mythological figures, and surround oval panels containing paintings of Bacchus.

64 The stucco decoration of the cupola of the Franciscan Church at Krosno, Poland, *c.* 1674, seems to be by a team of Milanese stuccoists. The dominant cherubs set into six splayed arms of interlocking stucco are an unusual feature.

Farthest north

While Scandinavia reflected mainstream European art, with a certain time-lag, England remained wedded to its own traditions until the eighteenth century.

65 French influence reached Sweden with the architect Nicodemus Tessin the younger, at the royal palace in Stockholm in the 1690s. Here, in the Charles XI Gallery the French sculptor Bernard Chauveau, a pupil of Girardon (whose work we have noted at Vaux-le-Vicomte, *see* pl. 41), and the Italian stuccoist Pietro Pagani, set out their figure groups with considerable skill. Ranged across the cornice and frieze, with a draped stucco curtain around the base of a bust of the Dowager Queen Hedwig Eleonora, a court of Diana assembles. The frieze contains trophies of the chase, and the corner cartouches elsewhere in the Gallery are also lively and competent.

66 Detail from the dining-room ceiling of Forde Abbey, Dorset, done for Edmund Prideaux, a leading supporter of Cromwell, *c.* 1655. English plaster, though coarser than Continental stucco, could achieve effects of some delicacy. This ceiling may be the work of the Devon plasterers Richard Abbot and his son John.

67 Coleshill, Berkshire, was built for George Pratt in the early 1650s to the design of his amateur-architect cousin, Sir Roger Pratt. Regrettably this splendid English house was totally demolished after a fire in 1952. The most likely author of the lavish, heavy plasterwork is John Grove the elder, Master Plasterer to the Office of Works.

68 Ceiling of the chapel of Belton House, Lincolnshire, *c.* 1687. The plasterer here was Edward Goudge, working in a style that is characteristic of its time, in which the figures and foliage, though lively in themselves, are subordinated to a conventional pattern of coffers and ribs.

69 Stucco and alabaster doorway of the palace of the Marquis de Dos Aguas at Valencia, 1740–4, by Hipólito Rovira Brocandel. The giants at either side, said to represent Flora and Fauna, but Michelangelesque in stature, emerge from a tumultuous luxuriance of plant and animal life. They empty water vessels, signifying the two rivers of Valencia but also making allusion to the name of the owner. The group of the Virgin and Child is by another hand. The effect is theatrical, and the line between sacred and profane a fine one.

IV

ROCOCO THE LINE OF BEAUTY

The Rococo exploited the many possibilities of asymmetry, and in so doing, it frequently attained levels of exquisite balance. S- and C-shaped curves in an infinite variety of forms – as waving vines, delicate scrolls, branches of sweeping acanthus and undulating mouldings – melted by imperceptible stages into each other. It was a complete abandonment of Renaissance formalism.

Germany

The idea of a natural order had gained in momentum during the late seventeenth century, when the princes of Church and State developed an architecture and decoration to follow on from the rich exuberance of the Baroque. The mixing of figures with heavy acanthus-scrolling which had characterized such important interiors as Fischer von Erlach's Mausoleum at Graz (*c.* 1687) gave way to delicate Berainesque ribbon-work, and to a free Rococo style introduced from France in the early eighteenth century. The chronology of this French introduction is unsettled. In France the change had started by 1700; in Germany it appears in the work at Nymphenburg of 1716–19, and continues until the late 1760s. During the early eighteenth century some stuccoists moved to Augsburg, not only to be near the undoubted centre for the diffusion of ornamental engravings, but to create liaisons with the many fresco-painters living there.

Fig. 11

Though German art was moving slowly away from that of Italy, the interpretation of motifs was at first sometimes gross and ill-conceived. The Duchesse d'Orléans, who had decorated various rooms in the Hôtel du Petit Luxembourg in Paris in the Rococo style, wrote from Paris in 1721 about the situation: ' . . . Germany not only imitates France but always does double what is done here'. Versailles was now the noble pattern, and many a princeling struggled to encompass some semblance of it in the *galeries*, staircases and succession of impressive state apartments in his Schloss. What these evolutionary processes required was a transition, whenever possible, to the style based on nature and the rendering of plants and shells. It allowed a tension and pattern within its seemingly abandoned asymmetry, and still conformed to its architectural setting.[1]

It found acceptance at the many courts,[2] encouraging patrons to set up their own groups of artists, removed from the restrictive attitudes of the guilds.[3] Status was achieved by a display of ostentation, which might require the spending of some three-quarters of available revenue on decoration. The work was distributed to carpenters, cabinet-makers, carvers, workers in stucco, wood, paint, precious and semi-precious metals, upholsterers and tapestry-weavers.

An Imperial Decree of 1731 directed to modernization of the German guild system allowed court artisans to be employed in greater numbers, and to achieve the status already accorded to the goldsmiths and painters.[4] Some centres of creativity such as Nuremberg, Augsburg and Ulm had

The Asams: dramatists of the supernatural

No artists came closer to conveying religious emotion in terms of sheer spectacle than Cosmas Damian and Egid Quirin Asam. Even today, to come upon one of their works in a small Bavarian church is to experience a shock of surprise and delight.

74 View into the dome of the abbey church of Weltenburg, completed by 1723. A detail from this composition is shown in pl. X. The effective blend of gilded stucco panels and frescoes draws attention to the advantages the Asams derived from their Roman training. At Weltenburg most of the sources of the light that illumines the dome, and indeed the whole church, are hidden, so that the effects are dark, mysterious and dramatic.

75 Nave ceiling of the abbey church of Rohr, built at about the same time as Weltenburg and also by the Asams. The whole of the vault is covered by a single framed space, but instead of the fresco which was probably intended, there is an ornate rendering of the monogram of the Virgin surrounded by areas of gold mosaic and stucco clouds.

76 The abbey church at Osterhofen, because of its size and wealth, could command important decorations, 1731–2, from the Asam brothers (*see also* pl. VIII). This symbolic group is of 'Faith trampling Evil underfoot'.

77 Egid Quirin Asam built the Munich church of St John Nepomuk (1733–46) as a private chapel. Beyond the cornice we look upwards to Cosmas Damian's fresco of the martyred saint, and above the altar to the stucco group representing the Trinity—God the Father, the Crucified Son and the Holy Spirit in the form of a dove.

The Asam's rivals

South Germany in the early and mid-eighteenth century was peculiarly rich in stuccoists of superb originality and accomplishment. Among the most outstanding was Johann Baptist Zimmermann, a stuccoist who worked usually in conjunction with his brother Dominikus, architect and fresco-painter.

78 Schloss Schleissheim, near Munich: stucco Atlantes from the Hall of Victories by Charles Dubut, finished before 1726. They form the upper part of the wall decoration of the hall, the lower two thirds consisting of gilded wood-carving. Dubut was a French stuccoist who came to the Munich court from service in Berlin.

79 Detail of the capitals and frieze below the undulating cornice of Günzburg Parish Church (1736–41), by the Zimmermann brothers. In its lightness, its flowing line and formal freedom one recognizes the transition from Baroque to Rococo.

80 Pilgrimage church of Steinhausen, Baden-Württenberg, 1731. Steinhausen is a simple domical canopy church that has been rendered one of the masterpieces of European decoration by the skill of the Zimmermanns. The whole ceiling is covered with an enveloping fresco by Johann Baptist, in which the Virgin as Queen of Heaven soars above in golden clouds to the sound of trumpets. Above the capitals, stucco Apostles by Dominikus gaze around or upwards at the painted groups of figures representing the Four Continents. In this view, looking east, is the Fountain of Life. The Garden of Eden is at the west end, along with Dominikus's signature as 'ARCHIT E STUCKADOR' and the date 1731. The church is flooded with light: stucco birds are set on the capitals and in the window-bays as if they had flown in from a blue frescoed sky, and white *putti* swing floral garlands to each other. It is a concept of great beauty.

The court of Munich

A succession of enlightened Electors helped to make Bavaria one of the centres of secular as well as religious art.

81 The elaborately decorated rooms in the Munich Residenz by the court architect François Cuvilliés are among the finest decorative achievements of the Bavarian Rococo. This view into the Grüne Galerie (1733) shows the snaking, gilded stuccoes with an effective use of background space.

82 Details from the ceiling of the Spiegelsaal ('Room of Mirrors') in the Amalienburg—a garden-pavilion in the park at Nymphenburg on the outskirts of Munich—by Johann Baptist Zimmermann, 1734–9. The Spiegelsaal is one of the loveliest interiors in Europe, with the Wittelsbach blue and silver represented in the blue background to the silvered frescoes. The modelling of the figure and its casual pose on the cornice, has all the effortless virtuosity of Rococo art at its peak.

The church in the meadow

The pilgrimage church of Die Wies ('the meadow') is perhaps Zimmermann's masterpiece, as captivating in its details as in its total effect (*see also* pls XII, XIII).

83-4 Elevation of the choir of Die Wies with (below) Zimmermann's original drawing, of about 1745: a composition of great complexity, with spaces seen through the arcades and openings above, but handled with superb control.

85 One of the Fathers of the Church (St Jerome), of wood covered with stucco in the nave of Die Wies. The sculptor was Anton Sturm of Füssen.

86 Abbey church of Our Lady at Ettal, 1744–62. The frescoed ceiling and a series of magnificent side-altars and balconies compete for attention with the pink, white and gilded organ screen, a combination of wood and stucco of surpassing elegance.

A British initiative

Virtually every style used in England from the Renaissance onwards had been imported into Britain from abroad. Neoclassicism alone could claim to owe as much to British artists as to those of any other country.

124 (previous page) Dining Room of Syon House, near London, 1761–9, designed by Robert Adam and executed by Joseph Rose (*see also* pl. XVI).

125 Roundel from Ardress House, near Portadown, Ireland, 1770s, showing 'Cupid bound with nymphs'. The stuccoist was Michael Stapleton of Dublin, one of Ireland's most outstanding decorators.

126 Detail from the library of Hatchlands, Surrey, 1758–61, designed by Robert Adam for Admiral Edward Boscawen.

127 Powerscourt House, Dublin, 1773. In Dublin one finds the same vocabulary as that used by Adam—scrolling arabesques and classical urns—adapted with equal facility by stuccoists such as James McCullagh, here assisted by Michael Reynolds.

128 Detail from the decoration of the Pavillon de Bagatelle, in the Bois de Boulogne, Paris, 1777. This series of finely modelled figures, bearing urns on their heads and lyres in their hands, can be seen as part of the French equivalent to the Adam style in England. Bagatelle, designed by François-Joseph Belanger, was commissioned and finished by the Comte d'Artois in sixty days—to win a bet with Marie Antoinette.

SEDLMAYR, Hans, *Österreichische Barockarchitektur 1690–1740*, Vienna 1930.
SIMONA, Luigi, *Artisti della Svizzera Italiana in Torino e Piemonte*, Zurich 1933.
*——*L'Arte dello Stucco nel Cantone Ticino, Pt. I. Il Sopraceneri*, Bellinzona 1938.
*——*L'Arte dello Stucco nel Cantone Ticino, Pt. II. Il Sottoceneri*, Bellinzona 1949.
STARING, A., 'Zeventiende-seuwe stucwerk in het oosten des lands' in *Nederlandsche Kunsthistorisch Jaarboek* (1948–9), pp. 316–40.
STÜRMER, Michael, 'An Economy of Delight: Court Artisans of the Eighteenth Century' in *The Business History Review*, LIII, No. 4 (Winter 1979), pp. 496–528.
TINTELNOT, Hans, *Die barocke Freskomalerei in Deutschland. Ihre Entwicklung und europäische Wirkung*, Munich 1951.
VOLK, Peter, *Rokoko plastik*, Munich 1981.
WAGNER, Helga, and Ursula PFISTERMEISTER, *Barock Festsäle in bayerischen Schlössern und Klöstern*, Munich 1974.
*WIENERROITHER, J. M., *Steirische Innendekoration von den ersten Deckengestaltungen italienischer Stukkateure im 16. Jahrhundert bis zum 18. Jahrhundert*, Dissertation, University of Graz (1952).
*WITTKOWER, Rudolf, (II) *Art and Architecture in Italy, 1600–1750*, 3rd edn., Harmondsworth 1973.
——*Studies in Italian Baroque* (ed. Margot Wittkower), London 1975.
YANU, J. 'Les stucateurs de Wessobrunn' in *Connaissance des Arts* (Jan. 1978), pp. 50–7.
ZENDRALLI, A. M., *Graubündner Baumeister und Stukkatoren in deutschen Landen zur Barock- und Rokokozeit*, Zurich 1930.
ZYKAN, J., 'Barocker Stuck und seine Pflege' in *Zeitschrift Deutsche Kunst- und Denkmalpflege* (1942–3), pp. 117 *ff.*

(ii) Lives of Artists
AGGHAZY, Maria G., 'Alcun lavori della bottega di Giovan Battista Barberini nell' Ungheria d'altri tempi' in *Arte Lombarda*, vol. 11, no. 2 (1966), pp. 163–8.
BASILE, E., *Le scolture e gli stucchi di Giacomo Serpotta*, Turin 1911.
BEARD, Geoffrey, 'Bagutti: a most ingenious artist' [Giovanni Bagutti of Rovio] in *Apollo*, 97 (May 1973), pp. 489–91.
BLAŽIČEK, Oldrich J., 'Giacomo Antonio Corbellini e la sua attività in Boemia' in *Arte Lombarda*, vol. II, no. 2 (1966), pp. 169–76.
BOECK, Wilhelm, *Joseph Anton Feuchtmayer*, Tübingen 1948.
BRAUNFELS, Wolfgang, *François de Cuvilliés . . .*, Würzburg 1938.
BÜCHNER, Wolfram, *Der Stukkator Johann Baptist Modler von Kosslarn, ein Meister des deutschen Rokoko*, Passau 1936.
*CARADENTE, G., *Giacomo Serpotta*, Turin 1967.
*CAVAROCCHI, F., 'I Lurago, Quali Stuccatori' in Arslan, E. (ed.), *Arte e Artisti dei Laghi Lombardi*, Como 1964, pp. 33–48.
——'Diego Francesco Carlone e Giacomo Antonio Corbellini nel tricentenario della nascita' in *Rivista di Como*, IV (1974), pp. 2–11.
CORNFORTH, John, 'The Franchini in England', *Country Life*, 147 (12 March 1970), pp. 634–6.
*DISCHINGER, Gabriele, *Johann und Joseph Schmuzer: Zwei Wessobrunner Barockbaumeister*, Sigmaringen 1977.
DONATI, U., 'Antonio Raggi' in *L'Urbe*, VI, ii (1941).
DÖRY, Ludwig Baron, *Johann Martin Hummel, ein unbekannter Stukkateur in der Rhön*, Fulda 1954.
*——'Donato Giuseppe Frisoni und Leopoldo Mattia Ratti' in *Arte Lombarda*, vol. 12, no. 2 (1967), pp. 127–38.
*DŮRAS, A., *Die Architekten familie Lurago*, Prague 1938.
ENGL, Franz, 'Die Stuckarbeiten Giovanni Battista Carlones in der St. Ägidiuskirche zu Vöcklabrück, in der Schlosskapelle zu Marbach, im Pfarrhof zu Ried und im Stifte Reichersberg' in *Arte Lombarda*, vol. 11, no. 2 (1966), pp. 149–54.
——'Die Stuckarbeiten Bartolomeo Carlones und Giovanni Manfredo Madernis im Stift St. Florian' in *Passauer Jahrbuch für Geschichte*, Jb. 11, Passau (1969), pp. 163–8.
FLEMING, John, and Hugh HONOUR, 'Giovanni Battista Maini' [1690–1752] in *Essays in the History of Art presented to Rudolf Wittkower* (eds. D. Fraser, H. Hibbard and Milton J. Lewine), II, London 1967, pp. 255–8.
FÖLLMAN, L., 'Dominicus Zimmermanns . . . Planung bei der Barockisierung der ehem. Klosterkirche zu Gutenzell, 1755–6 . . .' in *Das Münster*, 28, no. 5–6 (December 1975), pp. 296–9.
GAVAZZA, E., 'Del Barberini plasticatore lombardo' in *Arte Lombardo*, VII (1962), pp. 63–74.
GRIMSCHITZ, Bruno, *Johann Lucas von Hildebrandt*, Vienna, Munich 1959.
GÜNTHER, Erich, *Die Brüder Zimmermann*, Königsberg 1944.
GÜRTH, A., 'Lurago, Carlone und Dientzenhofer' in *Christliche Kunstblätter*, Bd. 97, Linz 1959, pp. 5 *ff.*
HANFSTAENGL, Erika, *Die Brüder Cosmas Damian und Egid Quirin Asam*, Munich, Berlin 1955.
HANSMANN, Wilfried, 'Zwei Entwürfe von Johann Adolf Biarelle im Wallraf-Richartz Museum' in *Wallraf-Richartz Jahrbuch*, 34 (1972), pp. 357–62.
HITCHCOCK, Henry-Russell, 'The Schmuzers and the Rococo Transformation of Medieval Churches in Bavaria' in *Art Bulletin*, 48 (1966), pp. 159–76.
——'The Brothers Asam and the Beginnings of the Bavarian Rococo Church Architecture: I. Through the early 1720s; II. From the early 1720s to the mid 1730s', in *Journal, Society of Architectural Historians*, 24 (1968), pp. 186–228; 25 (1969), pp. 3–48.
*——[II] *German Rococo: The Zimmermann Brothers*, London 1968.
HOFFMAN, Hans, *Der Stuckplastiker Giovanni Battista Barberini, 1625–91*, Augsburg 1928.
HOJER, Gerhard, 'Die frühe Figuralplastik Egid Quirin Asam', Dissertation, University of Munich (1967).
HOPPE, T., 'Elia Castellos Stuckdecken im Neubau in Salzburg' in *Österreichisches Zeitschrift für Denkmalpflege*, V (1951).
IRONSIDE, W. Dalton, 'Giacomo Serpotta (1656–1732), A Neglected Artist in Stucco' in *Royal Institute of British Architects, Journal*, vol. 41, ser. 5 (13 October 1954), pp. 1046–50.
*JÍRA, J., *Karel Lurago*, Prague 1922.
KÁSZONYI, A., *Andrea Bertinalli stukkátor es köre*, Budapest (Iparmüvészeti Museum, 1964).
*KEMPEN, Wilhelm von, *Der Stuckateur und Baumeister Giovanni Simonetti*, Berlin 1925.
KOSEL, K., *Hans Jörg Brix und der unbekannte Stuckbildhauer von Wettenhausen*, Weissenhorn 1970.
LAVIN, Irving, *Bernini and the Unity of the Visual Arts*, 2 vols., New York, Oxford 1981.
LENTINI, R. (ed.), *Le scolture e gli stucchi di Giacomo Serpotta*, Turin 1911.
MAGNI, M. C., 'Considerazioni su Gian Battista Barberini stuccatore lainese' in *Commun. Miscellanea di scritti in onore di Federico Frigerio*, Como 1964, pp. 309–29.
MARANGONI, M., *I Carloni*, Florence 1925.
*MARTINOLA, Giuseppe, 'L'Itinerario in Terra Tedesca dello Stuccatore Giovan Battista Clerici di Meride' in Arslan, E. (ed.), *Arte e Artisti dei Laghi Lombardi*, Como 1964, pp. 303–16.
*——'Documenti per lo stuccatore Agostino Silva' in *Bolletino storico della Svizzera Italiani* 1973, I.
MERTIN, Paul, 'Dominikus Zimmermann, sein Verhältnis zu Johann Jakob Herkomer und Füssen' in *Aus Alt-Füssen. Beilage zum Füssener Blatt*, 18, Nos. 1–4 (1957).
*MIES VAN DER ROHE, Waltraut, 'Franz Joseph Holzinger, ein österreichischer Stuckator und Stuckbildner des 18. Jahrhunderts', Dissertation, University of Munich (1945).
MOREL, Andreas F. A., *Andreas und Peter Anton Moosbrugger: Zur Stuck-dekoration des Rokoko in der Schweiz*, Berne 1973.
MORITZ, Hans-Karl, 'Profane und profanierte Stuckarbeiten Carlones in Passau' in *Passauer Jahrbuch für Geschichte*, 11, Passau (1969), pp. 24–5.
MUCHALL-VIEBROOK, Thomas, *Dominikus Zimmermann*, Leipzig 1912.
NISSER, W., *Daniel Anckermanns stuckateur*, Fornvännen 1939. (German stuccoist working in Sweden, mid-17th century.)
*NIZZOLA, S. G., and M. MAGNI [I] 'Una traccia per Francesco Silva, stuccatore ticinese' in *Arte Lombarda*, 37 (1972), pp. 86–94.
*——and M. MAGNI, [II] 'Agostino Silva da Morbio Inferiore' in *Arte Lombarda*, 40 (1974), pp. 110–29.
NORBERG-SCHULZ, C., *Kilian Ignaz Dientzenhofer e il barocco boemo*, Rome 1968.
PAUKER, W., 'Donate Felice von Allio und seine Tätigkeit im Stifte Klosterneuburg' in *Beiträge zur Baugeschichte des Stiftes Klosterneuburg*, I, Vienna-Leipzig 1907, p. 60.
*RAGGIO, Olga, 'Alessandro Algardi e gli stucchi di Villa Pamphili' in *Paragone*, no. 251 (1971), pp. 3–38.
REUTHER, H., *Die Kirchenbauten Balthasar Neumanns*, Berlin 1960.
SAUREN, H.-M., *Antonio Giuseppe Bossi, ein fränkischer Stukkator*, Würzburg 1932.
SCHNEIDER, F. R., *Die Bildhauerfamilie der Verhelst in München und Augsburg*, Auerbach 1937.
SCHNELL, Hugo, 'Die Scagliola-Arbeiten Dominikus Zimmermann' in *Zeitschrift des*

Deutschen Vereins für Kunstwissenschaft, 10 (1943).
SEDLMAYR, Hans, *Johann Bernhard Fischer von Erlach*, Vienna 1956.
*THON, Christina, *J. B. Zimmermann als Stukkator*, Munich 1977.
*VOLLMER, Eva C., *Der Wessobrunner Stukkator Franz Xaver Schmuzer*, Sigmaringen 1979.
*WITTKOWER, Rudolf, (I) *Gian Lorenzo Bernini: The Sculptor of the Roman Baroque*, 3rd edn., London 1981.
WOLF, F., 'Balthasar Modler, der bedeutendste Dekorkünstler Nierderbayerns' in *Ostbairische Grenzmarken*, V (1961), pp. 104–7.
——'Der Stuckateur Franz Xaver Feichtmayer' in *Zeitschrift des Historischen Vereins für Schwaben*, 59/60 (1969), pp. 251–69.
ZORZI, Giangiorgio, 'Alessandro Vittoria a Vicenza e lo scultore Lorenzo Rubini' in *Arte Veneta*, vol. 5 (1951), pp. 141–57.

F. Neoclassical and Later

BEARD, Geoffrey, 'The Rose Family of Plasterers' in *Apollo*, LXXXV, no. 62 (April 1967), pp. 266–77.
CARDIFF, Joseph A., *Early Stucco Houses*, New York, The Atlas Portland Cement Co., 1916.
FEIST, Peter H., 'Neo-Classicism and Gothic Revival at Wörlitz' in *Neoclassicismo*, Genoa 1973, pp. 31–46.
FISCHER and JIROUGH Company, Cleveland, *Catalogue of Interior and Exterior Decorative Ornament*, Cleveland (Ohio), ?1910.
JACOBSON & COMPANY, *A Book of Old English Design: 47 Plates of Historical English Ornament*, New York 1921.
KALJAZINA, N. V., 'Lepnoj dekor v žilom interere Peterburga pervoj četverti XVIII veka' in *Russkoe Iskusstvo*, Moscow 1974, pp. 109–18.
MEDICI-MALL, K., *Lorenz Schmid 1751–99, wessobrunner Altarbauer und Stukkateur*, Sigmaringen 1975.
PUTERS, A., *Vasalli et Gagini stucateurs italiens au pays de Liège*, Liège (1960).
QUINAN, J., 'Daniel Raynerd, stucco worker' in *Old-Time New England*, vol. 65, no. 3–4 (Winter 1975), pp. 1–21.
RALEY, Robert L., 'Early Maryland Plasterwork and Stuccowork' in *Society of Architectural Historians, Journal*, vol. 20, no. 3 (October 1961), pp. 131–5.

VI EXHIBITION CATALOGUES

Barock am Bodensee – Architektur, Bregenz, Künstlerhaus, 1962.
Barock am Bodensee – Plastik, Bregenz, Künstlerhaus, 1964.
L'arte del barocco in Boemia, Prague, National Gallery, 1966.
Barock in Deutschland – Residenzen, Berlin, 1966.
Barock in Nürnberg 1600–1750, Nuremberg, Germanisches Nationalmuseum, 1962.
Barock in Oberschwaben, Weingarten, 1963.
Bayerisches Rokoko, Munich, Bayerisches Nationalmuseum, 1946.
Bozzetti und Modelletti der Spätrenaissance und des Barock, Vienna, Kunsthistorisches Museum, 1937.
Johann Bernhard Fischer von Erlach, Graz, Vienna, Salzburg, 1956–7.
HOTZ, J. (ed.), *Katalog der Sammlung Eckert aus dem Nachlass Balthasar Neumanns im Mainfränkischen Museum Würzburg*, Würzburg, 1965.
Italienische Medaillen und Plaketten von der Frührenaissance zum Ende des Barock, Berlin, Staatliche Museen, 1966.
Linzer Stukkateure, Linz, Stadtmuseum, September–November 1973.
Kurfürst Clemens August, Brühl, 1961.
Kurfürst Max Emanuel, Bayern und Europa um 1700, Schloss Schleissheim, Munich, 1976.
Charles Le Brun, 1619–1690, peintre et dessinateur, Versailles, 1963.
Kurfürst Lothar Franz von Schönborn 1655–1729, Bamberg, Neue Residenz, 1955.
Mostra documentaria e iconografica di Palazzo Pitti e Giardino di Boboli, Florence, Archivo di Stato, 1960.
Balthasar Neumann in Baden-Württemberg, Stuttgart, Staatsgalerie, 1975.
Passavia sacra, Passau, 1975.
Jakob Prandtauer und sein Kunstkreis, Melk, Vienna, 1960.
Jakob Prandtauer, 1660–1726, der Baumeister des Österreichischen Barocks, Innsbruck, Museum Ferdinandeum, 1961.
Prinz Eugen als Freund der Künste und Wissenschaften, Vienna, 1963–4.
The Age of Rococo, Council of Europe, Munich, 1958.
The Romantic Movement, Council of Europe, London, 1959.
Johann Conrad Schlaun, 1695–1773, Münster Landesmuseum, 1973.
SCHONATH, Wilhelm (ed.), *250 Jahre Schloss Pommersfelden (1718–1968)*, Pommersfelden, 1968.
Die Vorarlberger Barockbaumeister, Einsiedeln and Bregenz, 1973.

Select Dictionary of Stuccoists and Plasterers

This list does not claim to be exhaustive. To provide a complete list of known stuccoists and plasterers would need more space than is available here. It does offer a preliminary guide to many important and representative figures, with an indication (again necessarily selective) of their commissions. For more detailed information, the reader is referred to the specialist publications noted in the bibliography.

Abbreviations
Short titles following the abbreviation *Lit* refer the reader to the Select Bibliography. The *Dizionario Biographico degli Italiano* (Rome, 1960 ff.) is abbreviated as *DBDI*; 'signed and dated' as 's. & d.'; 'destroyed' as 'dest.'.

ADAMI, Andrea (*fl.* 1702)
Of Lugano. Worked in several churches.
Lit: L. Brentani, *Antichi Maestri d'Arte e di Scuola Ticinesi*, III, pp. 174–5.

ADAMI, Giovanni Battista (*fl.* 1729–66)
Of Carone.
1729 TORICELLA, Parish Church, putti on altar.
c. **1750** VILLA LUGANESE, Parish Church, altar with inscription.
1766 VIRA E MEZZOVICO, Parish Church, altar.
Lit: Simona, II, p. 12.

AICHER, Thomas (*fl.* 1705)
c. **1705** ENSDORF, Abbey Church, crossing stucco. Remainder of work by M. T. and M. Ehamb (q.v.), 1708.
Lit: Hitchcock, *RASG*, p. 233, n. 15.

ALBUZIO, Giuseppe Antonio (*fl.* 1755–7)
1755–7 DÜSSELDORF, Schloss Benrath.
Lit: A. Klein, *Schloss Benrath* (1952).

ALFIERI, Hieronymus (1654–after 1720)
Worked mostly in Austria.
1684–7 KREMSMÜNSTER, Refectory, Library.
1701–2 VIENNA, Church of St Dorothea and the Deutschordenskirche.
Lit: Preimesberger, p. 327; Sailer, p. 62.

ALGARDI, Alessandro (1595–1654)
Born at Bologna, moved to Rome *c.* 1625. Also sculptor.
c. **1629** ROME, S. Silvestro al Quirinale, statues of SS. Mary and John the Evangelist.
Lit: Wittkower (I), pl. 162; Pope-Hennessy, p. 445.
1644–8 ROME, Villa Doria-Pamphili.
Lit: Raggio, pp. 3 *ff*.
c. **1648–50** ROME, S. Giovanni in Laterano, reliefs about aedicules.
c. **1650** ROME, S. Ignazio.

ALIPRANDI, Antonio (1654–after 1709)
Master stuccoist in 1680. Worked mostly in Austria for Johann Fischer von Erlach and Lukas von Hildebrandt.
1689 VIENNA, Church, Heiligenkreuz.
1700 VIENNA, Palais Harrach.
Lit: B. Grimschitz, *Wiener Barockpaläste*, Vienna 1947, p. 6.
Antonio's brother, Christoph Aliprandi, helped Giacomo Antonio Corbellini (q.v.), *c.* 1701–5.

ALLIO family
A Giovanni Battista Allio was born in Vienna in 1644, worked in Prague 1667–96. His son(?) Giovanni Battista Allio was born in 1706 and was active in Austria from 1720 to 1750. The date of his death is not recorded. His commissions included:
1720–2 NIEDERALTEICH, Abbey Church.
1724 KREMSMÜNSTER, Abbey Church.
1732 ZWETTL, Schloss Schwarzenau.
1736 and **1739** KLOSTERNEUBURG, Abbey Church.
1750 PASSAU, pulpit etc.
Lit: Guldan, pp. 179–82; Preimesberger, p. 331.

ALLIO, Paolo d' (*fl.* 1681–1722)
Worked in Austria and Germany. Assisted his relative G. B. Carlone at Passau. His commissions included:
1682 PASSAU, Cathedral.
1701–6 STRAUBING, Karmelitenkirche; RATTENBURG, Servitenkirche; SALZBURG, Kollegienkirche with D. F. Carlone (q.v.)
1708 SALZBURG, Schloss Klesheim, with D. F. Carlone.
1717 LAMBACH, Abbey Church.
1717 St Florian, Abbey Church.
1720 KREMSMÜNSTER, Abbey Church; AMBERG, Refectory, Paulankirche, altar in Mariahilfkirche; DOMMELSTADT, Parish Church.
Lit: Hermanin, I (Rome 1934), pp. 23–4; *DBDI*, I, pp. 414–15.

AMADEO, Giovanni Battista (*fl.* 1683)
One of a team working in Austria at Kremsmunster and St Florian, 1683.
Lit: Sailer, p. 66.
An Andrea Amadio of Lugano worked in Austria *c.* 1702.

ANCKERMANN, Daniel (*fl.* 1616–56)
A German stuccoist from Mecklenburg who worked in Sweden until 1656.
1616–23 DARGUN, Castle.
1643–53 SKOKLOSTER, equestrian monument of Herman Wrangel in Church.
1649–52 STRÄNGNÄS, Cathedral, funerary chapel (Admiral Gyllenhielm).
Lit: Karling, pls. 253–4.

ANDREA de Conti (*fl.* 1527–8)
1527–8 MANTUA, Palazzo del Tè. Assisted Niccolò da Milano and Primaticcio.
Lit: Verheyen, pp. 49, 128–30.

ANDRIOLI, Francesco (*fl.* 1724–49)
One of his relatives (?), Giovanni Andrioli, worked in Denmark. Francesco's commissions included:
1724–8 OTTOBEUREN, Abbey Church. Assisted Carlo Andrea Maini in the 'Amigoni-Zimmer'.
Lit: H. Schnell, *Ottobeuren*, Munich 1955, p. 14.
1733 HIPOLTSTEIN, Parish Church.
c. **1733** FREYSTADT, Spitalkirche.
1736 BÄRNAU, Paris Church. Assisted by Hieronymous Andrioli.
1737 Worked in EICHSTATT and NUREMBERG.
c. **1740** BAYREUTH, Schloss Eremitage. Worked with C. D. Bossi.
Lit: Guldan, pp. 195–8.

ANDROJ, Johann Kajetan (*fl.* 1718–46)
1718 At GRAZ.
1731 VORAU, Cathedral Library.
1735–40 WENIGZELL, Parish Church.
1739 ST LAMBRECH, Church, Prälatensaal.
1738–41 TAMSWEG, Church.
1746 TAMSWEG, Schloss.
Lit: Preimesberger, p. 343.

ANTHONI (Antoni), Giovanni (*fl.* 1663–77)
Of Mainz. Worked for a time in Sweden.
1663–4 SKOKLOSTER, Castle, three ceilings.
1668 DJURSHOLM, Castle.
Lit: Karling, p. 292, pls. 255–6.

ANTONI, Carlo (*fl.* 1731–51)
1702–51 At PRAGUE, occasionally with Michele Ignazio Palliardi.
1731 Maltese Nuns' Convent.
1735 Kanka-Haus.
1740 Palais Colloredo.
1744 STRAHOV, Klosterkirche.
Lit: Blažiček, 'Lombardische Stuckateure', pp. 125–6.

APPIANI, Jacopo (*fl.* 1724–36)
Son of Pietro Francesco Appiani (q.v.).
1724 WALDSASSEN, Abbey Library. Worked with Paolo Marazzi. *Stuckmarmor* of altars in Abbey Church.
1728 RHEINAU, Abbey Church.
1729–36 FÜRSTENFELDBRUCK, Abbey Church, main aisle, and marbling (1736).
Lit: Guldan, pp. 199–200.

APPIANI, Pietro Francesco (*fl.* 1703–24)
Died Regensburg 1724.
Known for his stuccoes at Fürstenfeldbruck, Nymphenburg, Freystadt, and worked in Munich. Assisted by his son Jacopo (q.v.), who continued his work at Fürstenfeldbruck.
1723 FÜRSTENFELDBRUCK, Abbey Church.
Lit: Guldan, pp. 201–2; Lieb (I), pp. 147, 163.

ARTARI family (late 17th–18th c.)
The history of this family, settled at Arogno in the Ticino towards the end of the sixteenth century, is complex. An extensive literature (summarized in *DBDI*, IV, 1962, pp. 351–2) assumes Giovanni Battista Artari was born in 1660, and his son Giuseppe in 1697. However, Arogno parish registers show that Giovanni Battista was born in 1664. Giuseppe was born in 1692 or 1700 (not in 1697 as stated in most sources). His brother Adalbertus was born in 1693. All three worked in England. There was also an Artari family at Bissone nearby.

ARTARI, Adalbertus (1693–1751)
Only his English work is known.
1724 SUTTON SCARSDALE, Derbyshire.
1725 DITCHLEY, Oxfordshire.
Lit: Beard (1975), p. 201.

ARTARI, Giovanni Battista (1664–after 1730)
Born at Arogno. As some sources (e.g. Bénézit, *Dictionnaire . . . des peintres . . .*, 1948) give a third Christian name of 'Alberti' to him, he has been confused with his son Adalbertus.
1707 FULDA, Cathedral. Worked with G. B. Genone. Four figures of Fathers of the Church.
1723–30 RASTATT, Cathedral; AGUISGRANA, Cathedral.
The stuccoist Alfonso Oldelli (1696–*c.* 1770), writing to Giovanni Oldelli at Meride, 2 July 1721, mentions the 'Signori Artari' working in England, and states that they were doing well and that he had a mind to join them.
Lit: *DBDI*, IV, p. 351; Hermanin, II, pp. 35, 42, 44; Martinola, *Lettere dai paesi* . . ., p. 117.

ARTARI, Giuseppe (1692/1700–1769/71)
Born at Arogno. The confusion in parish register entries prevents exact identification. Stated to have trained with his father and to have worked in Rome, Germany and Holland before reaching England. His wife 'Mary Gertrude' Artari received payment on his behalf on one occasion (1744–5). He left for Germany finally some time after 1760 to take up work again for the Elector of Cologne. He died there in 1769 (some sources give 1771).
Lit: Beard (1975), pp. 201–2.

English work
1720 TWICKENHAM, Octagon House.
1722–6 LONDON, St Martin-in-the-Fields.
1722–30 CAMBRIDGE, Senate House.
1723–4 LONDON, St Peter's, Vere Street.
1725 DITCHLEY, Oxfordshire.
1726 HOUGHTON HALL, Norfolk.
Lit: Beard (1981).
1729 LONDON, Cavendish Square.
1730–1 MOULSHAM HALL, Essex.
1736 CASTLE HOWARD, Yorkshire. Submitted plans for Temple of Four Winds decoration, given to Vassalli (q.v.).
1737 UPTON HOUSE, Banbury (s. & d.).
1737–8 TRENTHAM, Staffordshire (dest.).
1742 CASTLE HILL, Devon.
1743–4 WIMPOLE HALL, Cambridgeshire.
1744–5 OXFORD, Radcliffe Camera.
1756–60 RAGLEY HALL, Warwickshire.
For attributed work *see* Beard (1975), p. 202.

Continental work
1720–30 AACHEN, Cathedral, octagon of nave (removed 1870–3).
1729–32 FALKENLUST, BRÜHL, hunting-lodge.
Lit: *DBDI*, IV (1962), pp. 351–2.
1743–4 POPPELSDORF, Bonn. Worked with Carlo Pietro Morsegno and the brothers Castelli.
1748–61 BRÜHL, Schloss Augustusburg, various rooms, and staircase figures.
Lit: E. Renard and F. W. Metternich, *Schloss Brühl*, Berlin 1934.
1750–2 At MÜNSTER, and BONN, S. Clemente.

ASAM, Egid Quirin (1692–1750)
Son of frescoist Hans Georg Asam (1649–1711), and brother of the fresco-painter Cosmas Damian Asam (1686–1739). The list of his works is extensive. The more important of the ecclesiastical commissions include:
1717–22 INNSBRUCK, Jacobkirche.
1720 ALDERSBACH, Abbey Church. Frescoes by Cosmas Damian.
c. **1720** MICHELFELD, Abbey Church. Includes some dated work of 1716.
1721 WELTENBURG, Abbey Church, cove, decoration in dome, high altar (St George in silver).
1722–3 ROHR, Priory Church.
1731–2 OSTERHOFEN, Abbey Church, various altar figures, groups on and above high altar. Frescoes by Cosmas Damian.
1733–46 MUNICH, St John Nepomuk. Frescoes by Cosmas Damian.
Lit: Hitchcock, *RASG*, pp. 19–82; Lieb (I), pp. 162–9.

AUGUSTINI, Pietro (*fl.* 1749–51)
1749–51 FRIEDRICHSTHAL, Schloss, Orangery.
Lit: Baier-Schröcke, 'Lombardische Stuckateure', p. 109.

BAADER, Leonhard (*fl.* 1744)
Of Wessobrunn.
1733 GARMISCH, St Martin, choir, assisted Pallir Michael Schmidt.
Lit: Dischinger, p. 142.

BAADER (Bader), J. Georg (b. 1692)
Of Wessobrunn.
1714–15 MUNICH, Dreifaltigkeitskirche.
Lit: Hitchcock, *RASG*, p. 23.
1722–3 ROHR, Abbey Church, high altar, assisted E. Q. Asam.
Lit: Lieb (I), p. 163.
c. **1720** NYMPHENBURG, Badenburg, *stuckmarmor*.
c. **1725** SCHLOSS SCHLEISSHEIM, Munich. Worked in *stuckmarmor* with C. Dubut and J. B. Zimmermann in the Kammerkapelle.
Lit: L. Hager, *Schloss Schleissheim*, Königstein 1974, p. 7.

BAGUTTI, Antonio (*fl.* 1780)
Of Rovio.
1780 BESAZIO, Parish Church, altar cornice.
Lit: Simona, II, p. 69.

BAGUTTI, Bartolomeo (1720–92)
1784 ROVIO, Casa Bagutti, façade medallion.
Lit: Simona, II, p. 55.

BAGUTTI, Giovanni (1681–after 1730)
Born at Rovio in the Ticino, son of Bernard Bagutti and Angela Maria (*née* Falconi). In England by 1709. Must be regarded as the senior partner to Giuseppe Artari (q.v.). Daniel Defoe in his *Tour through the Whole Island of Great Britain* (1725) called him 'the finest artist in those particular works now in England'. He should not be confused with a painter Abbondio Bagutti, nor with Pietro Martire Bagutti of Bologna. No Continental commissions have been recorded for him.
1709–10 CASTLE HOWARD, Yorkshire, various rooms; Great Hall, stucco fireplace, *scagliola* niche.
1720 TWICKENHAM, Octagon House. Assisted by Giuseppe Artari.
1722–5 MEREWORTH, Kent.
Lit: Colen Campbell, *Vitruvius Britannicus*, London 1725, III, p. 3: 'the ornaments are executed by Signor Bagutti, a most ingenious artist'.
1722–6 LONDON, St Martin-in-the-Fields.
Lit: James Gibbs, *A Book of Architecture*, London 1728: 'the ceiling enrich'd with Fret-work by Signori Artari and Bagutti, the best fret-workers that ever came into England'; Beard (1981) citing accounts.
1723–4 LONDON, St Peter's, Vere Street.
Lit: Gibbs, op. cit., p. vii.
1725–6 CAMBRIDGE, Senate House, ceiling (replaced late 19th c.); alternative design by Bagutti in Gibbs Collection, Ashmolean Museum, Oxford, II, 63–4.
1730–1 MOULSHAM HALL, Essex (dest.).
c. **1732** MOOR PARK, Hertfordshire.

Date uncertain CASSIOBURY PARK, Hertfordshire (dest.).

Attributed works
c. **1714–15** WENTWORTH CASTLE, Yorkshire, staircase off Long Gallery.
1724 THE MYNDE, Hereford.
Lit: C. H. C. and M. I. Baker, *The Life and Circumstances of James Brydges, 1st Duke of Chandos*, London 1949, p. 272, n.
1731–5 CLANDON PARK, Surrey, hall, various rooms.

BAR, Jakob (1705–after 1751)
Worked in Switzerland at Engelberg (1735 and 1745) and St Urban (1749–51). Skilled in *stuckmarmor* techniques.
Lit: Lieb (II), p. 128.

BARBERINI, Giovanni Battista (*c.* 1625–91)
Born in the Como region, and reputedly a pupil in Rome of Ercole Ferrata. One of the most sought-after stuccoists in northern Italy. His work is to be found in Laino, Bellinzona (1661, 1687), Cremona, Bologna, Genoa, Mantua, Bergamo, Como (S. Cecilia 1687), and in Austria at Vienna (Servitenkirche, 1669), Kremsmünster and Linz.
Lit: H. Hoffmann, *Der Stuckplastiker, G. B. Barberini, 1625–91*, Augsburg 1928; and E. Gavazza in *Arte Lombarda* VII (1963), pp. 63 *ff*.

BARELLI, Agostino (*fl.* 1672–88)
From Bologna. Also architect.
1672–88 MUNICH, Theatinerkirche. Assisted Carlo Brentano-Moreti (q.v.).
Lit: Guldan, pp. 210–15, pls. 208–212.

BARUZZI, Giovanni (*fl.* 1636–57)
Of Caneggio. Visited Rome in 1636.
1657 SALORINO, Parish Church, Baroque-style stuccoes.
Lit: Simona, II, p. 63.

BECHTER, Konrad (*c.* 1795–after 1828)
Of Lingenau. At Donauschingen, *c.* 1728.
Lit: Lieb (II), p. 128.

BELLOTTO, Giovanni Battista (*fl.* 1652)
1652 LUGANO, S. Antonio.
Lit: G. Martinola in *Bollettino storico della Svizzera Italiana* (1942), pp. 59–72.

BELTRAMELLI, Domenico (*fl.* 1699–1717)
Of a Ticino family.
1699 BRA, SS. Trinità.
1710–11 CHERASCO, S. Maria del Popolo.
1712 SAVIGLIANO, five statues in convent church.
1713–17 CUNEO, S. Croce.
Lit: L. Simona, *Artisti della Svizzera italiana in Torino e Piemonte*, Zurich 1933, pp. 20, 49; Carboneri, p. 23, n. 4; 28–30, pls. 27–41.

BENTELE, Andreas (*fl.* 1749–51)
1749–51 ST URBAN, Abbey Church, Festsaal.
Lit: Lieb (II), p. 128.

BERCHTOLD, Johann Anton (*fl.* 1773–91)
Of Bludenz.
1773–5 BLUDENZ, Parish Church, high altar.
1790–1 Side altars and pulpit.
Lit: Lieb (II), p. 128.

BERNASCONI family
A family of stuccoists of this name was settled at Riva St Vitale in the Ticino. The following have been noted:

BERNASCONI, Alessandro (*fl.* 1695–1720)
Son of Constantino. Worked at Waldsassen Abbey Church at various dates between 1695 and 1720.

BERNASCONI, Bartolomeo (1755–1835)
Born Lugano. Worked in Genoa in the second half of the 18th c.
Lit: A. Cappellini, *Ville genovesi*, Genoa 1931, p. 34; *DBDI*, II, p. 327.

BERNASCONI, Bernato (*fl.* 1770–1820?)
Worked in England. C. R. Cockerell told the *Select Committee on Arts and their Connection with Manufactures* (1835–6) that 'a few artists still survived in Ireland and there remained in England a Mr Bernasconi till 1820 . . .'.
1770–84 CLAYDON, Buckinghamshire, hall and ballroom.
Lit: *Country Life*, 7 November 1952, pp. 1483–4.
c. **1790** OXFORD, New College Chapel, vault, organ screen and stalls.
Lit: *Country Life*, 19 April 1979, p. 1189.

BERNASCONI, Francesco (Francis) (1762–1841)
Said to have been the son of a Bartholomew Bernasconi (d. 1786). As well as being the most fashionable purveyor of Regency Gothic stucco in England, he had a large business in *scagliola* manufacture. He is recorded as working at 27 English houses. No Continental commissions known.
Lit: Beard (1981).

BERNASCONI, Giuseppe (*fl.* 1640–50)
Born at Riva St Vitale. His work for Francesco Borromini is noted in Ch. III, pp. 47–8.

BIANCHI, Battuti (*fl.* 1699)
1699 BRA, Confraternità della SS. Trinità. Received 535 lire.
Lit: Carboneri, p. 25.

BIANCHI, Bernado (*fl.* 1670)
1670 VALTICE, Czechoslovakia, Church. Worked with Giovanni Salvi, Domenico Moretti and Francesco Falcone.
Lit: Blažíček, 'Lombardische Stuckateure', p. 119.

BIANCHI, Isodoro (*fl.* 1642)
1642 TURIN, Castello del Valentino.
Lit: Carboneri, pp. 4, 25–6.

BOLLA, Giovanni Michele (1682–after 1724)
Born in Milan 1682; worked in Vienna, and under Santino Bussi (q.v.) at Dürnstein (1723–4).
Lit: Sailer, pp. 71 *f*.

BONARDI, Pietro Antonio (*fl.* 1680)
1680 CELLE, Stadtkirche.
Lit: Amerio, p. 97.

BOSSI family
An extensive family with the following stuccoists: Antonio, Augustin, Carlo Daldini, Ludovico and Materno. The principal, for his work at Würzburg and elsewhere, was Giuseppe Antonio.

BOSSI, Antonio Giuseppe (*c.* 1700–64)
Born Porto Ceresio, near Lugano. Died, insane, in Würzburg, 10 February 1764.
1727–8 OTTOBEUREN, Abbey Church, Salone Imperiale.
1733 WÜRZBURG, Cathedral, Schönborn chapel.
1734 Appointed court stuccoist by Prince-Bishop Friedrich Carl von Schönborn.
1735–57 WÜRZBURG, Residenz. Gartensaal, Weissersaal, Kaisersaal, Hofkirche and various rooms; his principal achievement and occupying most of his working life. The Weissersaal (1744) is particularly noteworthy.
Bossi also worked for the Schönborn family in Bamberg (1735), and at Schloss Seehof (1736), Schloss Werneck (1744–), Gaibach (1746–9), Veitshöchheim (1752–3), Münster Abbey Church (1749, 1753), Amorbach Parish Church (1753–4), Käppelle, Würzburg (1754). He also stuccoed Madonnas, and rooms in Würzburg houses.
Lit: H.-M. Sauren, *Antonio Giuseppe Bossi*, Dissertation, University of Würzburg (1932); *DBDI*, IV, pp. 287–9 (cites extensive bibliography).

BOSSI, Ludovico (*fl.* 1765)
Nephew of Antonio Giuseppe Bossi (q.v.). Worked at Würzburg Residenz in the mid-1760s, and was court stuccoist at Stuttgart. Another of Antonio Giuseppe's nephews, Augustin, assisted him.

BOSSI, Materno (1735–1802)
Born at Porto Ceresio, near Lugano, son of Antonio Giuseppe Bossi (q.v.). Worked at the Würzburg Residenz in the 1760s, where several of his moulds survive.
After **1777** WERNECK, Schlosskirche.
1774–84 EBRACH, Abbey Church.
Lit: R. Sedlmaier and R. Pfister, *Die fürstbischöfliche Residenz zu Würzburg*, Munich 1923.

BOTTA, Giacomo (*fl.* 1685–99)
Of Merebilia.
1685 In DRESDEN.
1699 LEIPZIG, Barfüsserkirche.
Lit: Baier-Schröcke, 'Lombardische Stuckateure', p. 109.

BRENNI, Francesco Giulio (*fl.* 1678–94)
Date of birth unknown, but of a family settled at Salorino, near Mendrisio, in the Ticino. Worked with his brother (?) Giovanni Battista (*fl.* 1676–1715) in Austria and Germany, at Salzburg, Ebrach,

Herrenchiemsee. Giovanni Battista signed the *stuckmarmor* high altar in the Jesuitenkirche in Bamberg 'Gio Batt. Breno, Italiano, Inventor et Fecit Anō 1701'.
A Giovanni Prospero Brenni worked at the Theatinerkirche, Munich (1672–88).
Lit: Guldan, pp. 207–8.

BRENNO, Giovanni Battista (*fl.* 1676–1714)
Of Mendrisio.
Worked extensively at Bamberg; churches of St Michael (1700–13) and St Martin (1714).
Lit: Döry, 'Italienischer Stuckateure 1650–1750', p. 132.

BRENTANI, Bernhardo (*fl.* 1683–5)
1683–4 FRIEDRICHSWERTH, Schloss, first floor rooms.
1685 EISENBERG, Schloss Chapel, worked with Bartolomeo Quadri.
Lit: Baier-Schröcke, 'Lombardische Stuckateure', p. 109.
A Carl Wilhelm Brentani (*fl.* 1749–50) worked at OPPURG, Schloss, Garden Room, in 1749–50.

BRENTANO-MORETI, Carlo (*fl.* 1662–88)
Worked in Nuremberg (1662), and at the Theatinerkirche in Munich (1672–88) with Agostino Barelli, Giovanni Prospero Brenni and Giovanni Niccolò Perti.
Lit: Guldan, pp. 210–15.

BRILLI, Giuseppe (*fl.* 1750, d. 1794)
1750 BRUHL, Schloss Augustusburg, staircase figures (Atlantes and caryatids). Assisted Giuseppe Artari (q.v.) and C. P. Morsegno.
Lit: E. Renard and F. W. Metternich, *Schloss Brühl*, Berlin 1934.

BUSSI, Cajetan (1692–after 1739)
Master 1724. Assisted his brother Santino Bussi at Klosterneuburg in 1738–9.
Lit: Sailer, p. 73.

BUSSI, Carlo Antonio (1658–1690)
Brother of Santino, a painter, in 1687 married a daughter of the Tencalla family of frescoists.
Lit: *Arte Lombarda* XI (1966), pp. 177–82.
Santino's son, Antonio Gaetano (b. 1689), assisted him at Klosterneuburg, and in 1728 worked for Count Esterhazy in the Church of the Trinity at Bratislava.
Lit: Preimesberger, pp. 328–9; Blažíček, p. 121; full account in *DBDI*, VI, p. 575.

BUSSI, Santino (1664–1736)
Born at Bissone in the Ticino, 28 August 1664, son of Giovanni Francesco Bussi and Anna Maria (*née* Pusterla). His family was celebrated for its artists and decorators. Bussi worked in Milan and then moved to Vienna to serve Prince Eugene of Savoy. In 1698 he became a Viennese citizen, and on 31 August 1714 was nominated court stuccoist. His commissions included:
1695–1704 VIENNA, Liechtenstein palace.
1698 VIENNA, Kinsky palace.
End 17th c. VALTICE, Bohemia, Franciscan church.
1702–5 VIENNA, Monastery Church of St Dorothea.
1704–6 VIENNA, Garden palace, Liechtenstein.
1713 SALZBURG, Mirabell palace.
1719 HIRSCHSTETTEN, Schloss Schwarzenberg.
1722–3 VIENNA, Upper Belvedere.
c. **1723** BRESLAU, Church of the Prince-Elector.
1722–3 DÜRNSTEIN, Abbey Church.
1724 MELK, Abbey, state apartments.
1728 VIENNA, Garden palace, Schwarzenberg.
c. **1728** VIENNA, Garden palace, Harrach.
c. **1730** VIENNA, Peterskirche.
c. **1736** KLOSTERNEUBERG, Abbey, state apartments.
He also contributed to many other smaller commissions.

CALLIGARI, Francesco (*fl.* late 17th c.)
PORZA, Parish Church.
Lit: Simona, II, p. 35.

CAMINADA, Michele (*fl.* 1710–47)
Worked principally in Germany at the castles of Stolberg and Blankenburg, the Andreaskirche at Hildesheim (1720) and the cathedral there (1725–34).
Lit: Amerio, p. 97.

CAMUZZI, Antonio (1615–1724)
Born at Montagnola. Trained in the studio of Antonio Casella. His brothers Francesco and Fabio assisted him.
1690 GENTILINO, S. Abbondio.
Worked also in BERGAMO, S. Maria Maggiore.
Lit: *DBDI*, XVII, p. 630–1.

CAMUZZI (Camuzio), Giovanni Pietro (*fl.* 1680–1718)
Date of birth unknown, died at Passau, 2 September 1718.
His commissions included:
1680 KATZENBERG, Castle chapel.
1682 GARSTEN, Abbey Church, with G. P. Camuzzi, G. B. and Bartolomeo Carlone, and Domenico Garon.
1690 PASSAU, Residenzplatz, Library of Bishop Philipp Graf von Lamberg.
1693–4 VILGERTSHOFEN, Abbey Church, with Peter Camuzzi, high altar.
1696 AMBERG, Salesian Church.
1699–1702 SUBEN, Convent Church.
1699–1702 SUBEN, Priory Church, refectory.
Lit: Guldan, pp. 216–17; *DBDI*, XVII, pp. 630–1.

CAMUZZI, Muzzio (1717–after 1759)
Born at Montagnola. Trained with his family. Date of death unknown.
1739 MUZZANO, Parish Church, Worked in the Bergamo area for twenty years or so.
1759 LUGANO, S. Rocco.
Lit: Simona, II, pp. 15, 26, 32; *DBDI*, XVII, pp. 630–1.

CAPRANI, Paolo (1752–1819)
Born at Laino, died in Spain. Worked near Madrid, and in the Como villages near Laino, such as Pellio Superiore.
Lit: Magni, p. 77.

CARCHANI, Giovanni Battista (*fl.* 1690–1)
1690–1 EISENBERG, Schloss chapel, assisted by Nicolaus Carchani.
Lit: Baier-Schröcke, 'Lombardische Stuckateure', p. 110.

CARLONE, Diego Francesco (1674–1750)
Born at Scaria, son of Giovanni Battista Carlone, and member of an extensive family of artists of various kinds. Died at Scaria, 25 June 1750. His brothers, Carlo Innocenzo and Bartolemeo, followed their father as stuccoists, and principally in the great commission of decorating the cathedral at Passau. In 1695 Diego went to Rome to complete his instruction, and to study the work of Ercole Ferrata. Apart from Passau (*see* Giovanni Battista Carlone), his commissions included:
1701 AMBERG, Salesian Convent; worked for Johann Fischer von Erlach.
1704–10 ST FLORIAN, Abbey Church.
1705 SALZBURG, Kollegienkirche, with Paolo d'Allio (q.v.).
1708 SALZBURG, Schloss Klesheim, with Paolo d'Allio.
1714–18 LUDWIGSBURG, Schloss, various rooms, including Hofkapelle (*c.* 1718), figures including 'Slaves'; East Mirror Gallery.
1715 ST FLORIAN, Abbey Church.
1719–20 KREMSMÜNSTER, Kaisersaal.
1723–5 WEINGARTEN, Abbey Church, figures on high and side altars.
1734 ANSBACH, Residenz, entrance hall, twelve reliefs; Festsaal, reliefs and other decoration.
Lit: A. Führer, *Residenz Ansbach*, Munich 1973.

CARLONE, Giovanni Battista (*fl.* 1668–1707)
Son of Pietro Francesco Carlone. Married Taddea Allio. Had as a pupil Paolo d'Allio (q.v.) who assisted him on many occasions. One of the most competent stuccoists at work in the late 17th c., and of the Comasque family resident at Scaria. His principal commissions included:
1673 SCHLOSS NEUBURG am Inn, near Passau.
1674 In Scaria for the birth of his son Diego Francesco Carlone (who became a stuccoist).
1675–7 PASSAU, Jesuitenkirche.
1677 PASSAU, Cathedral.
1679–82 GARSTEN, Abbey Church, working under his father until 1679, then assisted by Bartolomeo Carlone, his brother, and Giovanni Pietro Camuzzi.
1680–2 Further work at Passau Cathedral.
1683 SCHLIERBACH, Abbey Church.
1683–4 Work at Passau and Garsten.
1686 In Scaria for birth of his son Carlo Innocenzo Carlone.
1687 PASSAU, Chapel at Bishop's residence.
1688 GARTLEBERG, Pilgrimage Church.
1690 VOCKLABRUCK, St Aegidius-Kirche, various altars.
1692 PASSAU, Summer Schloss of Hackelberg, assisted by Giovanni Pietro Camuzzi.
1693 SCHLÄGL, Abbey Church, guest rooms.

1693 PASSAU, Cathedral, various altars.
1695–6 WALDSASSEN, Abbey Church.
1697 REGENSBURG, Cathedral.
1698 AMBERG, Salesian Church.
1699 DINGOLFING, Michaelskapelle in the Stadtpfarrkirche.
1701 STRAUBING, Carmelitenkirche.
1702–4 AMBERG, Maria Hilf, Pilgrimage Church.
Lit: Guldan, pp. 235–54.

CAROVERI, Giovanni (*fl.* 1677–86)
Of Milan.
1677–9 WEISSENFELS, Schloss chapel.
1679 EISENBERG, Residenz. Worked with Bartolomeo Quadri.
1681 LEIPZIG, Alte Börse.
1683 EISENBERG, Schloss chapel. Worked with Bartolomeo Quadri.
1683–4 FRIEDENSTEIN, Schloss.
c. **1686** Worked in Berlin under Giovanni Simonetti (q.v.).
Lit: Baier Schröcke, 'Lombardische Stuckateure', p. 110.

CASSELLI, Michele and Antonio (*fl.* 1616–18)
1616–18 NEUBERG-an-der-Donau, Jesuitenkirche.
Lit: Powell, p. 29.

CASTELLACIO, Santi (*fl.* 1641–4)
Of Rome. Worked with G. M. Sorrisi and G. B. Frisone (q.v.).
1641–4 FLORENCE, Palazzo Pitti, various rooms.
Lit: Campbell, *Pitti Palace*, pp. 90, 187, 229.

CASTELLI, Carlo Antonio (*fl.* 1709–34)
Of Lugano. Brother of Giovanni Pietro Castelli.
1709–10 ARNSTADT, Schloss.
1710 FRIEDENSTEIN, Schloss.
1711 FRIEDRICHSTHAL, Schloss, gallery.
1720–1 UNTERHAUS near GERA, Schloss.
1723–7 WÜRZBURG, Residenz.
1727–34 ALTENBURG, Residenz.
Lit: Baier-Schröcke, 'Lombardische Stuckateure', p. 110.

CASTELLI, Carlo Pietro (*fl.* 1728–50)
Worked with his brother Domenico at the castles of Clemenswerth, near Osnabrück (1736), and Poppelsdorf, and at the Palais Thurn und Taxis, Frankfurt-am-Main.
Lit: Amerio, p. 98.

CASTELLI, Ciprianus (*fl.* 1706–23)
Of Bissone.
1706 Worked with Bartolomeo Lucchese (q.v.) for two years, at the Saalfeld Residenz and the Meiningen Residenz.
1723 MANNHEIM, Schloss. Worked under Eugenio Castelli.
Lit: Baier-Schröcke, 'Lombardische Stuckateure', p. 110.

CASTELLI, Francesco (?–1712)
Of Melide. Worked principally in Venice at the Palazzo Merati.
Lit: Lorenzetti, p. 246.

CASTELLI, Giovanni Pietro (*fl.* 1695–1731)
Settled at Melide. His brother Carlo Antonio and his sons Carlo Pietro and Giovanni Domenico assisted him on several occasions.
1695 At COBURG.
1699 GODESBERG, St Michael.
1699–1700 BONN, Residenz.
1705 EISENACH, Schloss.
1709 At ALTENBURG.
1709 ARNSTADT, Schloss.
1710 At FRIEDENSTEIN.
1712 At FRIEDRICHSTHAL.
1717 ERFURT, Residenz.
1718 BIERBRICH, Schloss.
c. **1720** BONN, Poppelsdorf Schloss.
1723–6 WÜRZBURG, Juliusspital.
1724 WERNECK, Schloss.
1727–34 ALTENBURG, Schloss.
Lit: Baier-Schröcke, 'Lombardische Stuckateure', p. 111.

CATENAZZI, Antonio (*fl.* 1696–1725)
Of Mendrisio in the Ticino, and presumably one of the extensive family settled at Morbio Inferiore since the 16th century.
1969 ASTI, Castle.
1721–5 MENDRISIO, S. Giovanni.
Lit: G. Martinola, 'I Conventi di Mendrisio' in *Bolletino storico della Svizzera Italiana* (1945); Simona, II, pp. 61, 69.

CHRISTIAN, Anton (*fl.* 1716)
c. **1716** ST TRUDPERT, Black Forest, altar.
Lit: Powell, p. 89.

CHRISTIAN, Johann Josef (1706–77)
Of Riedlingen. One of the most skilled of woodcarvers, providing choir stalls, pulpits, confessionals, etc. Worked occasionally in stucco, assisted by his son Franz Josef Friedrich Christian.
1749 ZWIEFALTEN, Abbey Church, Ezekiel group, left altar. Assisted Johann Michael Feichtmayr (q.v.).
Lit: Lieb (I), pls, 92–3.
1754–64 OTTOBEUREN, Abbey Church, 'Baptism of Christ'; saints and angels, high altar. He and his son assisted Feichtmayr (as above).
Lit: Lieb (I), pls. 112–13.

CLARK, Thomas (*fl.* 1742–82)
Clark was one of the most successful English plasterers of the late 18th century. Based in London at Westminster, he was Master Plasterer to the Office of Works from 1752 until his death.
1745–60 HOLKHAM HALL, Norfolk, various rooms; Clark was working on the Saloon in 1753 and the Hall after 1759.
Lit: Holkham Archives, Building Accounts 8, 26.
1750 MILTON HOUSE, Northamptonshire.
1755 LONDON, Norfolk House.
Lit: West Sussex County Record Office, Duke of Norfolk's Archives; Victoria and Albert Museum, *Bulletin*, January 1966, pp. 1–11; D. FitzGerald, *The Norfolk House Music Room*, Victoria and Albert Museum publication, London, 1973, pp. 9–10.
c. **1782** LONDON, Somerset House.
Lit: Somerset House Accounts, British Architectural Library; Public Record Office, London, AO 3/244.

CLAYTON, Thomas (*fl.* 1710–72)
Plasterer. Clayton's birthplace is unrecorded, but it was probably in London. No record of his apprenticeship there has been traced. As far as is known, he worked entirely in Scotland, from about 1740 for at least 30 years.
1740 EDINBURGH, The Drum. Probably the Drawing Room.
Lit: Lennonxlove, Duke of Hamilton's Archives (Hamilton section, Box 127).
1747–57 BLAIR CASTLE, Perthshire.
Lit: Blair Castle Archives, 40 II D (4) 31–39. 40, III, 39–40 (letters).
1753–4 GLASGOW, St Andrew's Church.
Lit: Glasgow, City Minute Book, 9 March 1753; James Thomson, *History of St Andrew's Parish Church, Glasgow*, Glasgow 1905.

Attributed works
c. **1754** HOPETOUN HOUSE, West Lothian, Yellow and Red Drawing Rooms.
Lit: John Fleming, *Robert Adam and his Circle in Edinburgh and Rome*, London 1962, p. 332; *Country Life*, 12 January 1956.
c. **1756** DUMFRIES HOUSE, Ayrshire.
Lit: Sir John Stirling Maxwell, *Shrines and Homes of Scotland*, Edinburgh 1938, pp. 193–4.
c. **1760** YESTER HOUSE, East Lothian, Saloon.
Lit: Diary of Dr Richard Pococke, cited by John Swarbrick, *Robert Adam and his Brothers*, 1915, p. 220.
1771–2 EDINBURGH, 36 St Andrew Square.
Lit: British Library Add. Ms. 41133, *f*. 53; George Richardson, *Book of Ceilings*, London 1776, p. 4.

CLERICI, Giovanni Battista (*fl.* 1697–1730)
Of Meride in the Ticino. The training and work-itinerary of Clerici has been established for 1697–1730, and shows that a successful Ticino stuccoist might expect to work in many parts of Germany, Austria and Bohemia. Clerici worked at Baden, Bamberg, Würzburg, Kassel, Wabern, Lubeck, Wahlstorf, Hamburg, Strelitz, Berlin, Mannheim, Strasbourg, Karlsruhe, St Blasien and Darmstadt. His letters are in the State Archives at Bellinzona. The stuccoes by him (*c.* 1726) at the Abbey Church of St Peter in the Black Forest have been noted as 'dry proto-Rococo ornamentation'.
Lit: Martinola, *Clerici*, pp. 303–15; Hitchcock, *RASG*, p. 156.

COLLINS, Thomas (1735–1830)
A detailed account of the plasterer Collins' long career was prepared, 1965–6, by Col. J. H. Busby (copy British Architectural Library, London).
1765 WALCOT, Shropshire. Worked with William Wilton.
Lit: Bills at house.
1765–6 LONDON, 45 Berkeley Square.
Lit: India Office Library, Clive Papers.
1771 MILTON HOUSE, Northamptonshire, work under Sir William Chambers for Lord Fitzwilliam.
Lit: British Library, Add. Ms. 41133, 9 November 1771.
1773 LONDON, Piccadilly, Melbourne House.

Lit: British Library, Add. Ms. 41133, 14 August 1773.
1773 LONDON, 79 Stratton Street.
Lit: Northants County Records Office, Milton Archives, Vouchers, 114, and letter from Collins.
1777 PEPER HAROW, Surrey.
Lit: Bill of June 1777; *Country Life*, 26 December 1925.
1780 LONDON, Somerset House.
Lit: British Architectural Library, Library Accounts; Public Record Office, A.O. 3/1244.

COLOMBANI, Placido (*fl.* 1744–97)
Worked in England and Ireland.
1775 DOWNHILL, Co. Antrim, N. Ireland.
Lit: *Country Life*, 6 January 1950.
c. **1780** MOUNT CLARE, Surrey.
Lit: Christopher Hussey, *English Country Houses: Mid-Georgian*, London 1963, p. 240.
c. **1797** ICKWORTH, Suffolk.
Lit: *Country Life*, 7 November 1925; Hussey, op. cit., p. 240.

CONSIGLIO, Francesco (*fl.* 1734–9)
Worked in England.
1734 LYME PARK, Cheshire, staircase hall.
Lit: *Country Life*, 16 December, 1974.
1739 EUXTON HALL, Lancashire.
Lit: *Country Life*, 6 February 1975.

CONTI, Andrea de *see* **ANDREA de Conti**

CORBELLINI, Giacomo Antonio (?1674–1742)
Stuckmarmor specialist.
1707–8 POLNÁ, Church of the Ascension, vault.
1713–15 OSEK, Church of the Ascension.
1717–18 LUDWIGSBURG, Hofkapelle, *stuckmarmor*.
1719 WEINGARTEN, Abbey Church, high altar, *stuckmarmor*.
1731 ANSBACH, Residenz.
Lit: O. J. Blažíček, 'Giacomo Antonio Corbellini e la sua attività in Boemia' in *Arte Lombarda*, vol. II, no. 2 (1966), pp. 169–76; Hitchcock, *RASG*, p. 32; Powell, p. 98.
A 'Luca Corbellini', stuccoist of Lugano, worked in S. Antonio, Lugano, *c.* 1652.
Lit: G. Martinola in *Bollettino storico della Svizzera Italiana* (1942), pp. 59–72.

CORTESE, Giuseppe (*fl.* 1725–78)
Presumably of the family of stuccoists long settled at Mendrisio, near Lugano.
1739 NEWBURGH PRIORY, Yorkshire.
1745–52 STUDLEY ROYAL, Yorkshire (dest.).
1747–9 BRANDSBY HALL, Yorkshire.
c. **1750** GILLING CASTLE, Yorkshire, Great Hall.
1752 and **1757** ELEMORE HALL, Durham, various ceilings, one using the same Neptune motif as the ceiling at Lytham Hall, Lancashire.
1757 HARDWICK PARK, Durham, Garden Temple.
1762 BEVERLEY, Guildhall.
1769 BURTON CONSTABLE HALL, Yorkshire.
1772 KILNWICK HALL, Yorkshire.
Lit: Beard (1981), with references also to attributed work.

DALDINI, Carlo (*fl.* 1741–4)
Of Venice. Worked at Coburg, Bayreuth and the Ansbach Residenz.
Lit: Baier-Schröcke, 'Lombardische Stuckateure', p. 113.

DIRR, Johann Georg (*fl.* 1749–57)
Pupil of J. A. Feuchtmayer (q.v.). Worked with him at Birnau in stucco (1749) and wood (1757).

DONATI, Andrea (*fl.* 1799)
Of Lugano.
1799 ROVIO, SS. Vitale and Agata, columns to high altar, *stuckmarmor*.
Lit: Simona, II, p. 55.

DOOGOOD, Henry (*fl.* 1663–1707)
Doogood, with John Grove II, was a plasterer employed extensively by Sir Christopher Wren, and he worked at 32 London City churches. In 1700 he was made Master of the Worshipful Company of Plaisterers (London, Guildhall Library, Ms. 6122/3).
Lit: *The Wren Society* (1924–43), XII; Beard (1981).
1663 CAMBRIDGE, Pembroke College Chapel.
Lit: R. Willis and J. W. Clark, *The Architectural History of the University of Cambridge*, I, Cambridge 1887, p. 147; *The Wren Society* (1924–43), VI, pp. 27–9, pl. XI; N. Pevsner, *Cambridgeshire*, Harmondsworth, 1954, p. 27, pl. 58.
c. **1681** LONDON, St Mary Aldermary.
Lit: *The Wren Society* (1924–43), X, p. 13; A. E. Daniell, *London City Churches*, London 1896, p. 233.
1670–94 LONDON, City churches.
1682 and **1690** TUNBRIDGE WELLS, Kent, St Charles the Martyr, dated ceilings.
Lit: Marcus Whiffen, *Stuart and Georgian Churches*, London 1947, p. 97.

DUBUT, Charles (1687–1742)
French. Trained in Italy. Worked in Berlin under Andreas Schlüter.
c. **1720** NYMPHENBURG, Badenburg.
1723–4 SCHLOSS SCHLEISSHEIM, Munich. Assisted J. B. Zimmermann (q.v.) in various rooms.
Lit: L. Hager, *Schloss Schleissheim*, Berlin 1945; H. Wagner and U. Pfistermeister, *Barock Festsäle*, Munich 1974, p. 106.

EGELL, Paul (1691–1752)
Of Wessobrunn. A talented sculptor, but also versed in stucco techniques. Master of Ignaz Günther, and of the stuccoist J. A. Feuchtmayer (q.v.). The altar by him at Hildesheim is a fine work.

EHAMB, Mathias, Thomas and **Martin** (*fl.* 1708)
1708 ENSDORF, Abbey Church.
Lit: Hitchcock, *RASG*, p. 23.

ENGSTLER, Kaspar (*fl.* 1790)
1790 BLUDENZ, Parish Church, side altars.
Lit: Lieb (II), p. 128.

ERIKSSON, Nils (*fl.* 1670)
Swedish. Trained by Giovanni Anthoni and Hans Zauch.
c. **1670** SKOLOSTER, Castle.
Lit: Karling, p. 293.

FARILLO, Christoph (*fl.* 1687)
1687 EISENBERG, Schlosskapelle. Worked with Bartolomeo Quadri.
Lit: Baier-Schröcke, 'Lombardische Stuckateure', p. 111.

FEICHTMAYR family
Of Wessobrunn. Members include Franz Xaver (1705–65).
c. **1754** GUTENZELL, Nunnery Church, assisted by his son-in-law Jakob Rauch (q.v.).
Johann Michael (below) was Franz Xaver's brother. F.X. had a son, also Franz Xaver, who became assistant (1752) to J. B. Zimmermann. At Zimmermann's death in 1758, Franz Xaver II, then of Augsburg, married his widow and took over the stucco practice.

FEICHTMAYR, Johann Michael (1709–72)
Perhaps the most competent of the three Feichtmayrs. Among his extensive activities, the following important commissions may be noted:
c. **1739** DIESSEN, Priory Church, choir ceiling, assisted by J. G. Übelherr.
Lit: Lieb (I), pl. 69.
1747–8 ZWIEFALTEN, Abbey Church, stuccoes and *stuckmarmor*. Worked with Johann Josef Christian (q.v.).
Lit: Lieb (I), pls. 91–3.
1752 WÜRZBURG, Käppele.
1757–64 OTTOBEUREN, Abbey Church, stuccoes and *stuckmarmor*; certain figures by Johann Josef Christian (q.v.).
1764 VIERZENHEILIGEN, Pilgrimage Church, with J. G. Übelherr.

FERABOSCO, Pietro (*fl.* 1673–9)
He and his family were active in Austria and Bohemia in the second half of the 17th century.
1673–9 OSNABRÜCK, Castle.
Lit: Amerio, p. 98.

FEUCHTMAYER, Joseph Anton (1696–1770)
Born at Linz. Apparently not related to the Feichtmayr family (q.v.). Pupil of D. F. Carlone (q.v.) and Paul Egell of Wessobrunn. Died at Mummenhausen (monument by his pupil Johann Georg Dirr). Part of his youth spent in Salem where his father Joseph, a burgher of Augsburg, worked (1706). Working with Carlone brought him into touch with activity of the highest standard. His commissions included:
1728 ST PETER, Black Forest.
1730 EINSIEDELN, altars.
1741 MEERSBURG, Schloss chapel.
c. **1742** SALEM, Abbey.
Lit: Hempel, pl. 154A.
1748–58 BIRNAU AM BODENSEE, Pilgrimage Church, figures on altars.
Lit: Lieb (I), pls. 163, 168–71.
1760–70 ST GALLEN, Abbey Church, confessionals, choir-stalls.

Lit: Wilhelm Boeck, *Joseph Anton Feuchtmayer*, Tübingen 1948.

FEURSTEIN family (*fl.* 1st half 18th c.)
Of Schwarzenberg.
Stuccoists and altar-builders with at least six active members: Hans I, II, Johann Anton, Josef I, II, and Leopold.
Lit: Lieb (II), p. 128.

FISCHER, Josef (*fl.* 1729–32)
1729–32 Stams, Abbey Church. Worked with F. X. Feichtmayr I (q.v.).
Lit: Powell, p. 57.

FLOR, Johann Michael (*fl.* 1730–3)
1730–3 Altenburg, Abbey Library.
Lit: Hempel, p. 102, pl. 55.

FOSSATI, Giovanni Battista (*fl.* 1730–50)
Of Meride. Worked in Denmark with Giulio Guione, where both were court stuccoists.
1745– Christianborg, Castle.
1752 Amalienborg, Palace.
Lit: Grandjean, 'Stucateurs . . . en Danemark', p. 164.

FRANCHINI (Francini), Paul and Philip (*fl.* 1730–60)
Of Mendrisio. Active in England and Ireland, at Wallington, Lumley Castle, Fenham Hall, Carton, Castletown, etc.
Lit: Beard (1981); C. P. Curran, *Dublin Decorative Plasterwork*, London 1967; *Country Life*, 12 March 1970; 23 April 1970; 19 September 1974.

FRISONE, Giovanni Battista (*fl.* 1641–4)
Of Rome. Worked with G. M. Sorrisi and S. Castellaccio.
1641–4 Florence, Palazzo Pitti, Sala di Venere, Sala di Apollo, Sala di Giove, Sala di Saturno.
Lit: Campbell, *Pitti Palace*, pp. 90, 139, 187, 221, 229.

FRISONI, Donato Giuseppe (1683–1735)
Born at Laino, died at Ludwigsburg, 29 November 1735. An architect as well as stuccoist, as surveyor to the Duke of Württemberg (1717) he provided plans for various stages of building the basilica at Weingarten. He concerns us here by his involvement as a stuccoist (1709–) at the palace of Ludwigsburg, particularly in the *corps de logis* (the so-called Fürstenbau), and the banqueting house called Favorite. He also planned the new town of Ludwigsburg.
Lit: Döry, *Frisoni*.

GAGINI, Giovanni Francesco (*fl.* 1713–15)
Worked near Genoa with D. Beltramelli (q.v.).
1713–15 Cuneo, S. Croce.
Lit: Carboneri, pp. 21, 27–30.

GAGINO, Domenico (*fl.* 1679)
1679 Kromeriz, Bohemia. Worked with the architect Baldassare Fontana of Chiasso (1658–1738).

Lit: Blažiček, 'Lombardische stuckateure', p. 119.

GALLASINI, Andrea (*fl.* 1706)
Of Lugano.
1706 Meiningen, Residenz. Worked with Bartolomeo Lucchese (q.v.).
Lit: Baier-Schröcke, 'Lombardische Stuckateure', p. 111.

GALLI, Domenico (*fl.* 1644–75)
1645 Prague, Michna Palace.
1650–9 Prague, St Salvator in Clementinum. Worked with Carlo Lurago (q.v.).
Lit: Blažiček, 'Lombardische stuckateure', p. 118.

GALLI, Giovanni Maria (*fl.* 1685)
c. 1685 Vilna, SS. Peter and Paul. Worked with Pietro Peretti.
Lit: Karling, p. 298.

GALLI, Taddeo (*fl.* 1643–60)
1643 Graz, St Lambrecht.
1660 Seckau, Kaisersaal.
Lit: Preimesberger, pp. 338–9; Wieneroither, p. 20.

GAMBS, Lucius (1741–95) and **Alois** (c. 1765–?)
Lucius and his son Alois worked in Switzerland at Zurzach, Knonau, Hochsal, Laufenburg and Mettau.
Lit: Lieb (II), p. 129.

GAROVI, Pietro Antonio (*fl.* 1687)
1687 Feldsberg, Schloss, five rooms.
Lit: Sailer, p. 85.

GENONE, Giovanni Battista (*fl.* 1706)
Worked with Eugenio Castelli at a number of places in Germany.
Lit: Döry, 'Italienischer Stuckateure 1650–1750', pp. 136, 138, 140.

GIGL, Anton (d. 1757), **Johann Georg** (d. 1765), and **Matthaus** (*fl.* 1765–87).
Stuccoist family of Wessobrunn.
1757–9 Isny, St George, nave ceiling.
Lit: Bourke, p. 143.
c. **1765** St Gallen, Church and Library.
Lit: Boerlin, p. 124; Bourke, p. 190.
1765–6 Volders, St Carl Borromeo, nave ceiling, etc.
Lit: Bourke, p. 252.

GOUDGE (GOUGE), Edward (*fl.* late 17th and early 18th c.)
One of the most talented of English 'late Renaissance' plasterers. The letters of the architect Captain William Winde on occasion indicate works by Goudge.
1682–3 Combe Abbey, Warwickshire.
Lit: Bodleian Library, Oxford, Ms. Gough, Warwickshire, 1, 5 February 1682–3, 1 October 1683.
1684–8 Northampton, Sessions House.
1686 Thoresby House, Nottinghamshire.
Lit: Nottingham University Library, Pierrepont Archives, 4206.
c. **1688** Hampstead Marshall, Berkshire, ceiling (dest.).
Lit: Staffs County Record Office, Earl of Bradford's Archives, Winde Letters, Box 18/4; Geoffrey Beard in *Country Life*, 9 May 1952.

1688 Belton House, Lincolnshire.
Lit: letter of William Winde to Lady Bridgeman, Earl of Bradford's Archives; Geoffrey Beard in *Country Life*, 12 October 1951, p. 1157.
1688–90 Castle Bromwich, Warwickshire.
Lit: Geoffrey Beard in *Country Life*, 9 May 1952, citing Winde and Goudge letters in Earl of Bradford's Archives.
1691–2 Petworth, Sussex, chapel ceiling and hall of state.
Lit: West Sussex County Record Office, Egremont Archives, Richard Stiles' Account Rolls 1691–2.
1696–7 Chatsworth, Derbyshire, Gallery ceiling.
Lit: Chatsworth, Devonshire Archives, James Whildon's Account, 1685–99, pp. 121, 123, 125, 135; Francis Thompson, *A History of Chatsworth*, London 1949, pp. 56, 166–8.
Lit: Geoffrey Beard, 'The Beste Master in England', *National Trust Studies*, London 1979, pp. 20–7.

GREBER, Jakob (*fl.* 1747–51)
1749–51 St Urban, Abbey, Festsaal.
Lit: Lieb (II), p. 129.

GREUSSING, Johann (*fl.* 1713–41)
1720 Bregenz, Parish Church, high altar.
1740 Balgheim, altar.
1741 Donaueschingen, Parish Church, façade, gable-end.
Lit: Lieb (II), p. 129.

GROVE, John (*c.* 1610–1676), **John II** (?–1708)
During their careers both John Grove and his son became Master Plasterers to the Office of Works. John II succeeded his father in 1676. In 1657 Grove senior was Renter Warden of the Worshipful Company of Plaisterers. His son John II frequently worked with Henry Doogood (q.v.).
1661 Greenwich, Queen's House, East Bridge Room, ceiling.
Lit: *Survey of London*, XIV, 1937, pp. 72, 74, pls. 65–7.
1664–7 London, Clarendon House, Piccadilly (demolished 1683).
Lit: R. T. Gunther, *The Architecture of Sir Roger Pratt*, London 1928, p. 164.
1675 Cambridge, Emmanuel College Chapel.
Lit: R. Willis and J. W. Clark, *The Architecture and History of the University of Cambridge*, Cambridge 1887, II, pp. 703–9; *The Wren Society* (1924–43), V, pp. 29–31; N. Pevsner, *Cambridgeshire*, Harmondsworth 1943, p. 27.
1686–7 Cambridge, Trinity College, staircase to Library.
Lit: Willis and Clark, op. cit., pp. 533–51, esp. p. 540; *The Wren Society*, op. cit., pp. 32–44.

HAFFENECKER, Antonio (*fl.* 1748)
1748 Prestic, Bohemia, Church of the Assumption.
Lit: Cavarocchi, p. 43.

HANSCHE, Han (*fl.* 1677)
Belgian.

1677 CLEVE, Haus Hotzfeld.
Lit: Döry, 'Italienischer Stuckateure 1650–1750', p. 131.

HENDERSON, James (*fl. c.* 1755–87)
Plasterer, of York.
1762 YORK, Fairfax House, Castlegate.
Lit: Leeds, Yorkshire Archaeological Society Library, Newburgh Archives.
1765 HAREWOOD HOUSE, Yorkshire.
Lit: Cartwright Hall, Bradford, Danby Archives, Account Book II, p. 6; *Country Life*, 7 April 1966, p. 791.
1766–7 BARNSLEY, Yorkshire, Cannon Hall.
Lit: *York Georgian Society Report*, 1955–6, p. 59; Geoffrey Beard, *Cannon Hall Guidebook*, 1966.
1773 THIRSK HALL, Yorkshire.
Lit: North Riding of Yorkshire, County Record Office, Bills.

HENKEL, Andreas (*fl.* 1763)
Of Mindelheim.
1763 MUSSENHAUSEN, Church, aisle.
Lit: Eva Vollmer, *Franz Xaver Schmuzer*, Sigmaringen 1979, p. 70.

HENNICKE, Georg (?–1739)
Of Mainz.
1718 EBRACH, Abbey, guest range; ceiling of Kaisersaal; staircase.
Lit: Hitchcock, *RASG*, p. 7.
1722–3 POMMERSFELDEN, Schloss Weissenstein, Grotto (over stucco by Daniel Schenk, q.v.).

HOLZINGER, Franz Josef Ignaz (1691–1775)
1720 ST FLORIAN, Abbey Church. Assisted D. F. Carlone (q.v.).
Lit: Preimesberger, p. 337.
1722 METTEN, Abbey Church.
Lit: Hitchcock, *RASG*, p. 40.
c. **1730** OSTERHOFEN, Abbey Church.
Lit: Hitchcock, *RASG*, p. 65. W. Mies van der Rohe, *Holzinger*.

HOLZINGER, J. I. (*fl.* 1722–8)
c. **1728** NIEDERALTAICH, Abbey Church. Holzinger worked in the church over several years, assisting G. B. Allio (q.v.) and S. Allio.
Lit: Hitchcock, *RASG*, p. 178.

KLEBER, Johann (*fl.* 1736–49)
1736 SALZBURG, Schloss Leopoldskron.
1738 GNIGL (Salzburg), Parish Church.
1749 MITTERSILL (Salzburg), Parish Church.
Date unknown STUBAI (Tyrol) Parish Church, high altar, side altars.
Lit: Lieb (II), p. 129.

KOHLER, Peter (*fl.* 1821)
1821 KERNS, Switzerland, Parish Church, side altars.
Lit: Lieb (II), p. 129.

LAMONI, Felice (1745–1830)
Also architect. Of Muzzano.
c. **1770** MUZZANO, Parish Church.
Lit: Simona, II, p. 19.

LANCE, David (*fl.* 1691–1724)
Plasterer who in the late seventeenth century worked for Edward Goudge (q.v.). He was appointed Master Plasterer to the Office of Works in succession to John Grove II (q.v.) on 27 May 1708. His appointment was reconfirmed by George I on 27 May 1715.
Lit: *Calendar of Treasury Papers*, XXIX, Pt. 2, p. 102; *Survey of London*, XXIX, 1960, pp. 84 n., 100, XXVII, 1956, p. 33.

LANDES, Anton (?–1764)
Worked in the circle of J. G. Ubelherr (q.v.), T. Zöpf (q.v.), J. M. Merck (q.v.) and J. M. Feichtmayr (q.v.).
Lit: Eva Vollmer, *Franz Xaver Schmuzer*, Sigmaringen 1979, p. 39.

LEPORI, Antonio and Carlo (*fl.* 1750–60)
Of Origlio.
1758–9 ORIGLIO, Parish Church. (C. L.).
*c.***1760** MEZZOVICO, Parish Church. (A. L.).
Lit: Simona, II, pp. 12, 37.

LUCCHESE (Luchese), Carlo Domenico (*fl.* 1692–1724)
From Melide. Worked principally in Thuringia and Bavaria. His brother Bartolomeo was primarily an architect and a frescoist, and the two men often combined their respective skills.
1692 HILDBURGHAUSEN, Schloss, grotto and fountains.
1694 Unsuccessful bid for the stuccoing contract for Waldsassen Abbey Church, which was given to G. B. Carlone (q.v.).
1696–1700 SPEINSHART, Abbey Church. One of Carlo Domenico's best works; frescoes by Bartolomeo.
1697–1790 EHRENBURG, Schloss, state rooms, Schlosskapelle (1698).
Lit: H. Brunner, 'Die Bautätigkeit an Schloss Ehrenburg unter Herzog Albrecht' in *Jahrbuch der Coburger Landesstiftung*, 3 (1958), p. 175; H. Baier-Schröcke, 'Die Schlosskapelle der Ehrenburg . . . und ihre Stukkateure', ibid., pp. 195 *ff*.
1701 COBURG, Moritzkirche.
1704 GLÜCKSBURG, Schloss, for Herzog Heinrich von Römhild.
1704 SAALFELD, Residenz, state rooms, assisted by Giovanni Francesco Tattaro, Andrea Gallasini (q.v.) and Adolpho Piocha.
1713 GESTUNGHAUSEN, Parish Church and new buildings for the Duke of Coburg.
1717 Employed at Weingarten.
Lit: Baier-Schröcke, 'Lombardische Stuckateure'.

LUCHESE, Filiberto (*fl.* 1655)
Also an architect.
1655 LAMBACH, Abbey Church.
Lit: J. Schmidt in *Linzer Kunstchronik*, Linz 1952, III, pp. 87 *f*.

LUCCHESE (Luchese), Giovanni (*fl.* 1678–94)
Of the family of Bartolomeo and Carlo Domenico (q.v.), but his dates of birth and death, and his relationship to them, are unknown. Worked among the Italian 'colony' of artists in Vienna.
1694 WALDSASSEN, Abbey Church. Worked with G. B. Carlone.
Lit: A. Hajdecki, 'Die Dynasten-Familien der italienischen Bau- und Maurermeister der Barocke in Wien' in *Berichte und Mitteilungen des Altertums-Vereines zu Wien*, Bd. 39, Vienna 1906, pp. 31, 55, 83.

LUDWIG, Hans Georg (*fl.* 1742)
1742 SARNEN, Switzerland, St Martin, high altar.

LURAGO, Anselmo Martino (1701–65)
Born at Como, son of Giuseppe and Marsilia Carlone of Scaria. Not to be confused with his relative the stuccoist Giovanni Martino Lurago. Settled in Prague, and became a friend of the Austrian architect Kilian Ignaz Dientzenhofer. Best known for his stucco-work on the façade of the Kinsky Palace, Prague. Died in Prague, 29 November 1765.
Lit: Cavarocchi, pp. 42–4.

LURAGO, Carlo (1618–84)
Also architect. Born 14 October at Pellio Superiore, Valle Intelvi, son of Giovanni Antonio Lurago (1585–?), see below. Of an extensive family, of whom many nephews and cousins assisted him.
1638 PRAGUE, St Salvator, reconstructed and decorated for the Jesuits.
1650–9 PRAGUE, University, the Clementinum, assisted by his nephew Martino (1623–83), his cousin, Francesco Anselmo (1634–93) and Domenico Galli (q.v.).
1640–60 PRAGUE, Churches of the Knights of Malta and St Mary.
c. **1656** PRAGUE, Church of the Ascension; GLATZ, Jesuit college.
1657 PRAGUE, St Ivan. Also at Nachod, Pribam and Kladno.
1668 Appointed Imperial Architect by Count Wenzel of Thun-Hohenstein.
Lit: Jira; Důras; Cavarocchi; Blažiček in *Arte Lombarda*, 40 (1974), p. 152.

LURAGO, Francesco Anselmo (1665–?)
Worked in Prague and at Passau, with Carlo Lurago as architect. Collaborated on the construction of the church at Waldl in Bohemia.
Lit: Cavarocchi, p. 41.

LURAGO, Giovanni Antonio (1585–?)
Born at Pellio Superiore in Valle Intelvi and possibly the son or grandson of the architect Rocco Lurago (1525–90). Married Margherita Lurago di Anselmo. Had five sons, including Carlo Lurago (q.v.).
Lit: *Arte Lombarda*, X (1965), Pt. 2, p. 145.

LURAGO, (Giovanni) Antonio (1653–1727)
Born at Pellio Superiore. Settled in Prague. Also an architect, and friend of several artists including G. A. Corbellini (q.v.), and the architect-stuccoist D. G. Frisoni (q.v.).
Lit: Cavarocchi, p. 42.

LURAGO, Giovanni Martino (1719–75)
Born at Scarla, son of Giuseppe Tommaso Lurago and Maria Maddalena d'Allio (daughter of the stuccoist Paolo d'Allio). Active in Vienna and at Passau (Weisen-

kapelle), and Ludwigsburg. Died at Scarla, 15 December 1775.
Lit: Cavarocchi, p. 45.

MADERNI, Carlo (*fl.* late 17th c.)
Also an architect. Of Capolago. Worked at Capolago and Besazio.
Lit: Simona, II, p. 65.

MADERNI, Francesco (*fl.* 1676–84)
Worked in Austria.
1676–84 ALTÖTTING, Jesuitenkirche. Assisted G. B. Brenni and D. Martinetti, under the supervision of the architect Enrico Zuccalli.
Lit: R. A. L. Paulus, *Der Baumeister Henrico Zuccalli . . .*, Strasbourg 1912, p. 22.

MAINI, Andreas (*fl.* 1706–29)
1706–8 MEININGEN, Residenz. Worked for 2 years under the architect Bartolomeo Lucchese.
1715 OTTOBEUREN, Kaisersaal, Library. Assisted the Zimmermanns.
1715 At SCHWERIN-MECKLENBURG.
1728 GLÜCKSTADT, Wasmerchen Palace.
1729 OTTOBEUREN, Abbey Church.
Lit: Baier-Schröcke, 'Lombardische Stuckateure', p. 112.

MANSFIELD, Isaac (*fl.* before 1697–1739)
Plasterer, possibly born at Derby. His father Samuel was also a plasterer and was living at Derby at the time of his death.
Lit: *Surtees Society*, 102, p. 186; York Reference Library, Sessions Book, 1728–44; C. H. C. Baker and M. I. Collins, *The Life and Circumstances of James Brydges, 1st Duke of Chandos*, London 1949, p. 199.
1710 CASTLE HOWARD, Yorkshire. Assisted Giovanni Bagutti (q.v.).
Lit: Castle Howard Archives, Building Books.
1712–24 LONDON, St George's Church, Hanover Square.
Lit: H. M. Colvin, 'Fifty New Churches' in *Architectural Review*, March 1950, p. 196.
1714–28 LONDON, St John's Church, Westminster.
Lit: ibid.
1720–30 LONDON, St George's Church, Bloomsbury.
Lit: Colvin, loc. cit.
1720–21 LONDON, Burlington House.
Lit: Chatsworth, Burlington account book.
1721 CHICHELEY HALL, Buckinghamshire, hall and staircase.
Lit: Joan D. Tanner, 'The Building of Chicheley Hall', *Records of Bucks.*, XVIII, Pt. I (1961); *Country Life*, 20 February 1975, p. 437.
1721 LANGLEYS, Essex.
Lit: Essex County Record Office, Samuel Tufnell's accounts; *Connoisseur*, December 1957, p. 211.
1723–9 LONDON, Christchurch, Spitalfields.
Lit: Colvin, loc. cit.; Lambeth Palace Library, Ms. 2703.
1725–9 HOUGHTON HALL, Norfolk, work other than that by Giuseppe Artari (q.v.).
Lit: Cambridge University Library, Cholmondeley (Houghton) Archives, Vouchers, 1725, Account Book 40/1 (entry dated 19 July 1729).
1725 GOODWOOD, Sussex.
Lit: West Sussex County Record Office, Goodwood Archives, G. 121/1/107.
1725 BLENHEIM PALACE, Oxfordshire, Long Library, Chapel.
Lit: Blenheim Ms., E47, Hawksmoor to Sarah, Duchess of Marlborough, 23 December 1725, cited by David Green, *Blenheim Palace*, London 1951, pp. 310–11.

MARGETTS, Henry (*fl.* 1684–1704)
Plasterer who worked on Office of Works contracts under John Grove II (q.v.).
1684 EAST HATTLEY, Cambridgeshire.
Lit: Castle Howard Archives, Executor's Accounts, Sir George Downing.
1690 LONDON, Kensington Palace, outworks, stables.
Lit: Public Record Office, London, Works, 19/48/1, *f.* 108.
?–**1695** CHATSWORTH, Derbyshire.
Lit: Francis Thompson, *A History of Chatsworth*, London 1949, pp. 36, 59, 67.
c. **1700** KIVETON, Yorkshire. Received £372 16s od.
Lit: Leeds, Yorkshire Archaeological Society, Duke of Leeds Mss., Box 33.

MARINALE, Pietro (*fl.* 1598)
1598 FREIBERG, Cathedral, tomb of the Prince-Bishop.
Lit: Baier-Schröcke, 'Lombardische Stuckateure', p. 112.

MARTIN, Edward (*fl.* 1648–99)
Plasterer.
1671–81 LONDON, St Nicholas Cole Abbey, Queenhithe.
Lit: *The Wren Society* (1924–43), X, p. 73.
1678 ARBURY, Warwickshire, chapel ceiling.
Lit: Warwickshire County Record Office, Ms., CR 136/B24/3451; *The Wren Society*, op. cit., p. 22; Beard (1975), p. 142.
1682 BURGHLEY HOUSE, Northamptonshire.
Lit: Child's Bank, London, Exeter Bank Account, entries dated 1 July 1682, 14 February 1682–3.

MERCK, Johann Michael (1714–84)
1745 POTSDAM, Stadtschloss. Worked also at Sans Souci.
1765 At RAISTING.
1772 PAHL, St Laurence.
1772–3 At UNTERHAUSEN.
1775–9 ROTT, Parish Church.
1780 At RIECHLING.
Lit: W. Kurth, *Sanssouci*, pp. 157, 173; Eva Vollmer, *Franz Xaver Schmuzer*, Sigmaringen 1979.

MÉTIVIER, Joseph (*fl.* 1780–9)
Lived in Paris (Rue Nôtre-Dame de Nazareth).
1767 PARIS, Hôtel d'Uzès.
1780 PARIS, Hôtel Gouthière.
1789 PARIS, Boulevard Saint-Denis, Petit and Métivier tenements.
Lit: M. Gallet, *Paris Domestic Architecture of the 18th Century*, London 1972, p. 176.

MINETTI family (*fl.* 1702–20)
The family included Abondio, Carlo Antonio, Francesco, Gabriele and Johann Baptist. All four worked mostly in Thuringia, together except as noted.
1702 GREIZ, Stadtkirche (C.A.M.).
1703 ARNSTADT, Schloss (F. and A.).
1704 SCHANDAU, Parish Church (G.).
1708 WEISSENFELS, Schloss Augustusburg.
1709 EISENACH, Schloss, Pleasure Garden, Grotto. Worked with G. P. Castelli (q.v.).
1710 SONDERHAUSEN, Schloss (A. and F.).
1712–3 GOTHA, Schloss Friedenstein, gallery and four other rooms (F. and A.).
1716–9 QUERFURT, Schlosskirche (F. and A.).
1717–9 ZERBST, Schlosskapelle (F. and A.).
Lit: Baier-Schröcke, 'Lombardische Stuckateure', p. 112.

MOLA, Gasparo (1686–1746)
1725 OTTOBEUREN, Abbey Church.
1729–32 OCHSENHAUSEN, Abbey Church.
1732 WIBLINGEN, Abbey Church.
Lit: Döry, 'Italienischer Stuckateure 1650–1750', p. 141.

MOOSBRUGGER, Andreas (1722–87) and **Peter Anton** (1732–1806)
Both born at Schoppernau, sons of Franz Josef Moosbrugger. The brothers worked together on the commissions below except as noted. Their extensive works are recorded by A. F. A. Morel.
1758–61 TETTANG, Neue Schloss (A.).
Lit: Lieb (II), pl. 152.
1767 WÄDENSWIL, Parish Church, decoration of pulpit, (P.A.).
Lit: Lieb (II), pl. 146.
1774 WIL, 'Rudenzburg'.
Lit: Lieb (II), pl. 155.
1776–8 BERNHARDZELL, Parish Church, Rococo cartouches, framing to frescoes (P.A.).
Lit: Lieb (II), pl. 132.
1777 TEUFEN, Reformed Church.
Lit: Lieb (II), pl. 147.
1782 HERISAU, Parish Church.
Lit: Lieb (II), pls. 145, 151.
1789 MUOTATHAL, Parish Church, Rococo cartouches, balcony-front motifs.
Lit: Lieb (II), pl. 130.

MOOSBRUGGER, Hieronymus (1803–?)
1845–6 VIENNA, Niederösterreichisches Landhaus, conference room, walls and ceiling.
Lit: Lieb (II), pl. 174.

MOOSBRUGGER, Johann Josef (1771–1841)
Son of Andreas Moosbrugger (q.v.).
1808–12 GERSAU, Parish Church, choir (restrained Neoclassical decoration).
Lit: Lieb (II), pl. 172.
1809 WILLISAU, Parish Church.
Lit: Lieb (II), pl. 173.

The family also included the following stuccoists:
Johann Kaspar (1815–67), son of Josef Anton I (below); Johann Michael (1767–1831), son of Peter Anton II; Josef I (1701–69); Josef II (1703–41); Joseph III (1811–79); Joseph IV (?–1889); Josef Anton I (1764–1831); Josef Anton II (?–1877); Josef Leo (1774–1811); Josef

Leopold (1811–?); Josef Simon (1774–1831); Kaspar I (1656–1723); Kaspar II (1821–?); Leopold (*fl.* 1714–19); Michael (*fl.* 1725–55); Peter (*fl.* 1826–45); Peter Anton I (1732–1806); Peter Anton II (*fl.* 1801–46); Willibald (1813–44).
Lit: A. F. A Morel, *Andreas und Peter Anton Moosbrugger*, Berne 1973; Lieb (II), pp. 130–3, 141.

MORISI, Giuseppe Antonio (*fl.* 1776)
Worked with Johann Schmuzer.
1776 OBERMARCHTAL, Abbey Church, choir.
Lit: Eva Vollmer, *Franz Xaver Schmuzer*, Sigmaringen 1979, p. 106; Lieb (I), p. 145.

MUTTONE family (*fl.* 1667–1720)
The family included Carlo, Francesco Cristoforo (1670–1726), Giacomo and Niccolò.
c. **1667** GENOA, S. Luca; S. Croce; façade stuccoes, interiors. (C.M.)
Lit: Gavazza, pls. 60–1.
1679–80 GENOA, Palazzo Rosso. (G.M.)
Lit: Gavazza, p. 69.
1679 At CARIGNANO. (N.M.)
Lit: Gavazza, p. 69.
1719–20 AMBERG, Paulanerkirche. (F.C.)
Francesco Cristoforo also worked at Waldsassen (1695–8), with G. B. Carlone (q.v.).
Lit: Gavazza, p. 66; Guldan, p. 267.

NEGRI, Bernardo and **Giovanni** (*fl.* 19th c.)
Of Serocca d'Agno. Worked at Agno, Vezio (1830), Caslano, etc.
Lit: Simona, II, p. 27.

NICCOLÒ da Milano (*fl.* 1527–8)
Sometimes known as Brizio.
1527–8 MANTUA, Palazzo del Tè, Sala dei Venti, stucco festoons; Sala delle Aquile; assisted by Primaticcio and Andrea de Conti.
Lit: Verheyen, pp. 49, 114, 119, 121.

OLDELLI family (*fl.* late 17th–18th c.)
Of Meride in the Ticino.
The family was much-travelled, and also the recipient of an extensive correspondence (in the cantonal archives at Bellinzona) from other stuccoists abroad. Alfonso Oldelli (1696–*c.* 1770) led the family; Nazaro Oldelli was in Genoa (1667–71), and Bergamo (1685); Giovanni Antonio visited Rome in 1694, and Giuseppe and Gerolamo were at Mantua and Novara in the 1780s.
Lit: G. Martinola (ed.), *Lettere dai paesi transalpini degli artisti di Meride*, Bellinzona 1963.

PATROLI (*fl.* late 18th c.)
'An Italian artist of great ingenuity', long employed at Claydon, Buckinghamshire, under the supervision of the Rose family (q.v.) of plasterers. Perhaps to be identified with the 'Signor Pedrola' who worked with Joseph Rose senior (q.v.) at Ormsby Hall, Lincolnshire (*c.* 1755).
Lit: N. Lipscomb, *Buckinghamshire*, I, p. 186; N. Pevsner and John Harris, *Buildings of England: Lincolnshire*, Harmondsworth 1964, p. 370.

PEDROZZI, Johann Baptist (1710–78)
Born at Lugano, died in Berlin.
1728 OTTOBEUREN, Abbey Church. Worked under A. G. Bossi (q.v.).
1735 WÜRZBURG, Residenz. Worked under A. G. Bossi.
Pedrozzi was in charge of work at the Eremitage, Bayreuth, in 1750–64, and (with C. J. Sartori) at Sans Souci in Potsdam in 1764. He was in addition skilled at working *stuckmarmor*.
Lit: Baier-Schröcke, 'Lombardische Stuckateure', pp. 112–3.

PERGHOFER, Benno (*fl.* 1727)
c. **1727** MITTENWALD, Kreuzkapelle.

PERINETTI, Jacopo (*fl.* 1672–1716)
1672–3 WALDECK, Schloss Louisenthal, with P. M. Ferabosco.
1687 HANOVER, Leineschloss.
1690 SALZDAHLUM, Lustschloss.
1691 WOLFENBÜTTEL, Residenz.
1696 LUNEBURG, Palais d'Olbreuse.
1705–? BLANKENBURG, Schloss.
1716 DORSTADT, Klosterkirche.
Lit: Döry, 'Italienischer Stuckateure 1650–1750', p. 132.

PERRITT, Thomas (1710–59)
Trained by his father, Perritt dominated plastering in Yorkshire until his death in 1759. He was made a Freeman of York in 1737–8, and took his first apprentice, Joseph Rose senior (q.v.) in 1738.
Lit: York Reference Library, Skaife Mss., *Yorkshire Parish Register Society*, II, p. 196; *Surtees Society*, 102, p. 246; York, Borthwick Institute, Wills and Administrations, 13 December 1759.
1738–53 RABY CASTLE, Durham, various work, some in company with Joseph Rose senior (q.v.).
Lit: *Country Life*, 1 January 1970, pp. 20–1.
1741–7 LEEDS, Temple Newsam House, Long Gallery, Library and other principal rooms.
Lit: Leeds Archives Dept., Temple Newsam Archives, EA 12/10.
1744 YORK, Assembly Rooms.
Lit: York Reference Library, York Assembly Room Minute Book, entries as above, and similar ones for 5 June 1751, 12 July 1753.
1745 DONCASTER, Mansion House.
Lit: James Paine, *Plans, Elevations, Sections and other Ornaments of the Mansion House at Doncaster*, 1751.
1749 KILNWICK HALL, Yorkshire.
Lit: Edward Ingram, *Leaves from a Family Tree*, 1952.

Attributed work

c. **1740** NOSTELL PRIORY, Yorkshire, Dining Room, Music Room, North and South Staircase, medallions similar to those at Temple Newsam.
Lit: *Country Life*, 23 May 1952, pp. 1573–4.

PFLAUDER, P. (*fl.* 1782)
SALZBURG, Residenz, Marcus-Sitticus-Saal.
Lit: F. Martin, *Die Salzburger Residenz*, Augsburg 1928, pl. 12.

PLAZOL, Domenico (*fl.* 1690–8)
1690–8 At KLOSTERNEUBURG.
Lit: W. Pauker, *Das Stift Klosterneuburg*, Vienna 1935, p. 38.

POLLACK, Leopold (1751–1806)
1790–3 MILAN, Villa Reale-Belgiojoso, Sala da Ballo.
Lit: C. L. V. Meeks, *Italian Architecture, 1750–1914*, London 1966, p. 96.

POZZI, Francesco (*fl.* 1752–4)
Of Castel S. Pietro.
1752–4 OBERMARCHTAL, Abbey, refectory.
Lit: Eva Vollmer, *Franz Xaver Schmuzer*, Sigmaringen 1979, p. 73.

POZZI, Giuseppe (*fl.* 1759)
Of Castel S. Pietro. Worked in the church of his own village. Related to Francesco (above), and to Domenico (1744–96), a frescoist.
Lit: Simona, II, p. 61.

PRIMATICCIO, Francesco (1504–70)
Painter and stuccoist.
1527–30 MANTUA, Palazzo del Tè. Sala degli Stucchi and other rooms.
Lit: Barocchi, p. 209; Verheyen, pp. 124–5.
1530–1 MANTUA, Palazzo Ducale.
Lit: Verheyen, p. 124.
1531–2 FONTAINEBLEAU. Primaticcio arrived in France 23 March 1531. Assisted Rosso Fiorentino (1494–1540).
Lit: Johnson, pp. 9–18.

QUADRI, Bernardo (*fl.* 1688–1710) and Martino (*fl.* 1730–1)
Of a large stuccoist family settled at Lugano. Bernardo worked in south Germany, particularly at Bayreuth. Martino worked in England as an assistant to Francesco Vassalli. He is mentioned in the receipts for work at Towneley Hall, Burnley, Lancashire (1730).
Lit: Döry, 'Italienischer Stuckateure 1650–1750', p. 133.

QUADRIO, Antonio (*fl.* 1710–26)
Worked in Poland and Hungary.
1710 VARAŽDIN, Church of the Jesuits.
1710–11 LEPOGIAVA, Convent Library.
1712–26 ZAGREB, St Catherine.
1718 LEPOGIAVA, Parish Church.
1720 CERJE TUŽNO, Chapel of St Anthony.
Lit: *Arte Lombarda*, XI (1966), p. 144.

RAGGI, Antonio (1624–86)
Born at Vico Morcote, Como. Also a sculptor. Worked under Algardi, and then as Bernini's pupil from the late 1640s to *c.* 1670. His work included:
c. **1650** ROME, SS. Domenico e Sisto, *Noli me Tangere*.
c. **1652** PARIS, Notre-Dame, *Virgin and Child*.
1653 ROME, S. Maria sopra Minerva, tomb of Cardinal Pimentel, figure of *Charity*.
1655–9 ROME, S. Maria del Popolo, 'large part of the decoration'.
1656 ROME, Vatican, Sala Ducale, entrance decoration.
1658–64 ROME, St Peter's, assistance on the *Glory*.
1660–1 ROME, Castel Gandolfo, Church.

1660–7 ROME, S. Agnese in Piazza Navona, *The Death of S. Cecilia*.
1661–3 SIENA, Cathedral, Chigi Chapel (1662), statues of *S. Bernardino* and *Catherine of Siena*.
Lit: Pope-Hennessy, p. 438.
1662–5 ROME, S. Andrea al Quirinale, figures and decoration.
1667–70 ROME, Ponte S. Angelo, *Angel with the Column*.
1672–85 ROME, Il Gesù, nave and dome stuccoes; worked with the painter G. B. Gaulli, called Baciccia.
Lit: Donati; Enggass, pp. 3–74; Thieme-Becker, XXVII, p. 566; Wittkower I; Wittkower II, pp. 182 *ff*.

RAUCH, Jakob (*fl.* 1752–6)
Son-in-law of F. X. Feichtmayr (1705–65).
1752–3 ETTAL, Abbey Church.
1756 GUTENZELL, Nunnery Church. Assisted F. X. Feichtmayr (q.v.).
Lit: Bourke, pp. 151–2; Eva Vollmer, *Franz Xaver Schmuzer*, Sigmaringen 1979, pl. 131.

REALI, Michele and **Sebastiano** (*fl.* 1770–9)
Brothers, of Cadro in the Ticino.
1770–9 CADRO, Parish Church, nave, vault: inscription 'Opus Michaelis Reali ex pietate. 1779'; altars.
Lit: Simona, II, p. 41.

RETTI, Donato Riccardo (1687–1741)
Of Laino. A relative of the important architect (also from Laino) Donato Giuseppe Frisoni (q.v.).
1717 LUDWIGSBURG, Schloss. Worked with the Carlone family.
1725–6 MANNHEIM, Residenz.
1728–30 Worked at Schwarzach, Frauenalb and Attlingen.
Lit: Hitchcock, *RASG*, p. 31; Döry, 'Italienischer Stuckateure 1650–1750', p. 140.

ROBERTS, Thomas (1711–71)
Plasterer with an extensive practice in the Oxford area. Collaborated with the Danish stuccoist Charles Stanley (q.v.).
1738 OXFORD, Magdalen College, decorated colonnade of New Buildings.
Lit: W. G. Hiscock, *A Christ Church Miscellany*, Oxford 1946, pp. 68–71.
1742 OXFORD, St John's College, ceiling of the Senior Common Room.
Lit: Royal Commission on Historical Monuments, *Survey of City of Oxford*, 1939, p. 106.
1744 OXFORD, Radcliffe Camera. Assisted Charles Stanley (q.v.) to do eight ceilings.
Lit: *Oxford Historical Society*, 1958, XIII (1953–4).
c. **1745** KIRTLINGTON PARK, Oxfordshire. May have worked here with Charles Stanley. The *Aesop's Fables* medallions are a prominent feature.
1749, 1760 DITCHLEY, Oxfordshire, Dining Room.
Lit: Oxford County Record Office, Dillon Archives, I/p/3 ab, am.
1750 OXFORD, All Souls, Codrington Library.
Lit: Hiscock, op. cit., p. 71.
1752–62 OXFORD, Christ Church Library.
Lit: Hiscock, op. cit.
1756 OXFORD, Queen's College, Library ceiling. Roberts added 'new ornaments in the Oval Space in the Middle and the Compartments at the Ends'.
Lit: J. R. Magrath, *Queen's College*, Oxford 1914, p. 20.
1764 ROUSHAM, Oxfordshire, Great Parlour and other work.
Lit: *Country Life*, 24 May 1946, p. 949; Christopher Hussey, *English Country Houses: Early Georgian*, London 1965, p. 160.

ROSE family (*fl.* 1740–99)
The two principal members, of the four occupied in plastering, were Joseph Rose senior (*c.* 1723–80), apprentice of Thomas Perritt (q.v.), and Joseph Rose junior (1745–99). Joseph Rose senior had an extensive practice in the north of England, mainly for the architect James Paine. After Joseph Rose junior's return from Italy, *c.* 1766, the family firm set up in London, and turned increasingly to Neoclassical work for Robert Adam. They worked at almost every Adam commission, as well as doing some plastering for the architects Sir William Chambers and James Wyatt.
Lit: Beard, 1975 and 1981 (lists of commissions).

ROSSI, Carlo (*fl.* 1718–30)
Presumably son of Giovanni Domenico (below). Noted at the Residenzschloss of Braunschweig, the castle at Blankenburg, the church of St Andrea, and the cathedral at Hildesheim.
Lit: Amerio, p. 99.

ROSSI, Domenico Egidio (*fl.* 1704–7)
Also architect.
1704–7 RASTATT, Schloss. Ahnensaal.
Lit: Döry, 'Italienischer Stuckateure 1650–1750', pp. 133–4.

ROSSI, Giovanni Domenico (*fl.* 1654–74)
Worked in Bohemia and Germany.
c. **1654** NACHOD, Castle.
1661 CROTTORF, Schloss.
1668 TRIER, Cathedral, west choir.
1669–70 EHRENBREITSTEIN, Schloss.
1674 OSNABRÜCK, Schloss.
Lit: Blažiček, 'Lombardische Stuckateure', p. 118; Döry, 'Italienischer Stuckateure 1650–1750', p. 131.

SALVESTRINI, Cosimo (*fl.* 1644–7)
Pupil of Francesco Curradi.
1644–7 FLORENCE, Palazzo Pitti, Sala di Marte (assisted Pietro da Cortona); attributed, Sala di Venere, Medici portraits.
Lit: Campbell, p. 201, pls. 22–5; Thieme-Becker, XXIX, p. 360.

SARTORI, Carl Joseph (1709–60)
1749 POTSDAM, Stadtschloss, Marmorgalerie.
c. 1760 POTSDAM, Sans Souci.
Lit: W. Kurth, *Sanssouci*, pp. 157, 167.

SCAROLA, Giuseppe (*fl.* 1729–35)
c. **1734** NAPLES, S. Michele.
Lit: Blunt, *Baroque . . . in Naples*, p. 117, n.

SCHAIDHAUF, Thomas (*fl.* 1761–3)
1761–3 FÜRSTENFELD, Abbey Church, Apostle statues.
Lit: Lieb (I), p. 147.

SCHLAG, Jakob (*fl.* 1676–8)
1676–8 At KLOSTERNEUBURG.
Lit: W. Pauker, *Das Stift Klosterneuburg*, Vienna 1935, p. 59.

SCHENK, Daniel (d. 1737)
Of Bayreuth. Appointed 'Hofstuccateur' by Lothar Franz von Schönborn, who also sent him to Vienna.
1714–23 POMMERSFELDEN, Schloss Weissenstein, Gallery (1714), staircase (1714), Marmorsaal (1717), various first floor rooms. The quality of the stucco decoration varies, and some must be the work of assistants.
The Garden Grotto (1722–3) which included stucco-work by Schenk, was done under the direction of Georg Hennicke (q.v.) as 'Grottierer'.

SCHMID, Johann Maximus (*fl.* 1751–86)
Of Gaisboith. Married Maria Veronika Schmuzer, 6 February 1769.
Lit: Eva Vollmer, *Franz Xaver Schmuzer*, Sigmaringen 1979, p. 117.

SCHMUZER family (*fl.* 17th–18th c.)
Of Wessobrunn. The principal members and their relationships are as follows:

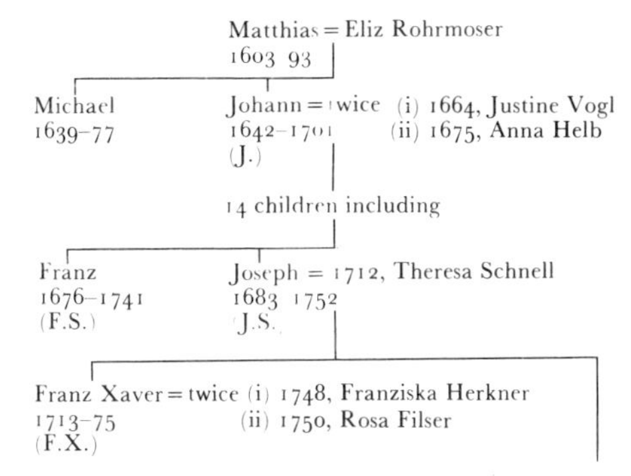

Matthias Schmuzer early involved his family with the Vorarlberg architect Michael Thumb, and carried out stucco-work for him on many occasions.
Matthias' son Michael worked for Thumb at the Jesuitenkirche, Lucerne (1672). His other son Johann acted as architect at 24 commissions documented, and at 7 attributed. Johann's son Joseph continued the architectural work, and Joseph's daughter, Maria, married the stuccoist J. G. Übelherr (q.v.).
Lit: Dischinger; Hitchcock I, pp. 159 *ff*.; Hitchcock, *RASG*, Ch. III; Lieb (I), pp. 12, 99–103, 128; Lieb (II), pp. 55 *ff*.; Vollmer (for Franz Xaver Schmuzer).

The family's commissions included:
1680 WESSOBRUNN, Abbey, guest range. (Later work, *c.* 1735, J.S.)
Lit: Lieb (II); Dischinger.
1689–92 OBERMARCHTAL, Abbey Church. Stucco contract 7 May 1689. (J.S. assisted by F.X., who completed after his father's death in 1701)
1687–92 VILGERTSHOFEN, Pilgrimage Church. (J.)
1691–1701 FRIEDRICHSHAFEN (formerly known as Hofen), Priory Church. Original stucco survives under the galleries. Rebuilt and re-stuccoed March–November 1950, simplified, but according to original plan.
Lit: Kosel (1969), pp. 101 *ff.*
1702 IRSEE, Abbey Church, ceiling, aisle and gallery (J.S.), restored 1978–80.
1707–9 RHEINAU, Abbey Church, ceilings. (F.S., frescoes by F. A. Giorgioli)
Lit: Landolt, pp. 49–52.
1718–19 WEINGARTEN, Abbey Church. Stucco contract 31 March 1718 (F.S., frescoes C. D. Asam).
Lit: Spahr; Dischinger (1977), pp. 49 *ff.*, 108 *ff.*; Lieb (I), pp. 145–6.
1718–19 WEISSENAU, Abbey Church, ceilings.
Lit: Hitchcock, *RASG*, p. 132.
c. **1720** DONAUWÖRTH, Heiligkreuz, Abbey Church (J.S.), ceiling, high altar (F.S.), 1724, s. & d.
Lit: Hitchcock, *RASG*, p. 133; Dischinger, p. 136.
1727–9 IRSEE, Prälatur, stair-hall. (J.S.)
Lit: Hitchcock, *RASG*, p. 135, pls. 131–2.
1732–3 GARMISCH, Parish Church. (J.S.)
Lit: Dischinger, p. 184.
1732–4 OBERAMMERGAU, Parish Church, ceilings, walls, pulpit figures, sacristry. (J.S. and F.S.)
1738–9 MITTENWALD, Parish Church. (F.S.)
Lit: Dischinger, p. 143.
1740–3 WEINGARTEN, Abbey, North quadrangle. (J.S. and F.X.)
Lit: Dischinger, pp. 151–2.
c. **1749–50** STEINGADEN, Abbey Church, Rococo-style framing to frescoes by J. G. Bergmüller (one s. & d. 1751). (F.S.)
1745–51 ETTAL, Abbey Church (Joseph). Assisted by his son-in-law J. G. Übelherr. Some gilded stucco.
Lit: Hitchcock, *RASG*, pp. 145–9; Dischinger, p. 166.

By Franz Xaver Schmuzer
1750 OSTERZELL, Parish Church.
1751–2 PETZENHAUSEN, Parish Church.
1751–2 MARIA STEINBACH, Pilgrimage Church.
1754–5 SCHMIECHEN, Pilgrimage Church.
1754–5 HEINRICHSHOFEN, St Andreas.
1760 SEEKIRCH, Parish Church.
1763 OBERMARCHTAL, Abbey Church, South pavilions.
1766 AMMERHOF, Parish Church.
Lit: Eva Vollmer, *Franz Xaver Schmuzer*, Sigmaringen 1979, p. 95, list of work (includes 15 attributed commissions).

SCHNEGG, Johann (*fl.* 1779)
c. **1779** WIBLINGEN, Abbey Church, group, *Sending Forth the Apostles.*
Lit: Powell, p. 127.

SCHWARZMANN, Jakob (*fl.* 1754–70)
1754–61 SCHUSSENRIED, Abbey, *putti* and Apostles.
Lit: Powell, p. 115.
c. **1770** WEINGARTEN, guest range, rooms 119, 127, 129, etc.
Lit: Eva Vollmer, *Franz Xaver Schmuzer*, Sigmaringen 1979, pp. 32, 114.

SERENA, Francesco Leone (*fl.* 1700–after 1729)
Born at Arogno in the Ticino, 5 November 1700. One of three sons of Domenico Serena and Giulia Cozzi. There is no record of his death.
Lit: A. Leinhard-Riva, *Armoriale Ticinese*, 1945; Registers at Arogno, 6 November 1700; 28 October 1774. Serena worked in England, but he is said to have also worked at Ottobeuren Abbey Church (presumably under J. A. Feuchtmayer, q.v.), and at the Landhaus, Innsbruck.
Lit: Thieme-Becker; Hugo Schnell, *Ottobeuren*, Munich 1950, pp. 22.
1725 DITCHLEY PARK, Oxfordshire. Worked with Giuseppe Artari (q.v.) and Francesco Vassalli (q.v.).
Lit: Oxford County Record Office, Dillon Mss, I/p/3h.
1729 LONDON, Cavendish Square. Received £30 for a knot-work ceiling done for James Brydges, 1st Duke of Chandos.
Lit: C. H. C. and M. I. Baker, *The Life and Circumstances of James Brydges*, London 1949, pp. 199, 277 n.

SERODINE, Giovanni Battista (1587–1626)
Of Ascona. Brother of the painter Giovanni Serodine (1594–1631). Responsible for the stucco decoration on the façades of the Casa Borrani, Ascona, and the Casa Rusca, Locarno.
Lit: Simona, I, pp. 8–9.

SERPOTTA, Giacomo (1656–1732)
Worked chiefly in Palermo (Oratory of S. Zita) and Agrigento (S. Spirito, 1693–5), Sicily. He was assisted occasionally by his illegitimate son, Procopio, who decorated the Oratory of S. Caterina, Palermo (1720–5).
Lit: Caradente; Blunt, *Sicilian Baroque.*

SILVA family
Of Morbio Inferiore in the Ticino. The family had at least four stuccoist-sculptor members: Francesco (see below), his son Agostino (1620–1706), Gian Francesco (1668–1737) and Carlo Antonio Benedetto (1705–88).
Lit: Martinola, II; Nizzola and Magni, II, Simona, II, p. 49.

SILVA, Francesco (1560–?)
1595–1613 MORBIO INFERIORE, S. Maria dei Miracoli.
1604– VARESE, Sacro Monte, Chiesa dell' Immacolata.
1610–16 LORETO, Santuario della Casa.
1616 FABRIANO, Cathedral.
1621–4 COLDERIO, Madonna del Carmelo.
1624 COMO, Cathedral; OSSUCCIO, S. Giuliano.
Lit: Nizzola and Magni, I; Brentani, II, pp. 141, 146.

SIMONETTI, Giovanni (1652–1716)
Born at Roverebo, died in Berlin. Held the post (from 1682) of 'Hofstuckateur' and 'Hofbaumeister' to the Elector Frederick William of Brandenburg. His commissions included:
c. **1670** PRAGUE, Czernin Palace.
1680 BRESLAU, Cathedral, Elisabeth Chapel.
1686 LEIPZIG, Börse.
1697–8 BERLIN, Schloss Oranienburg, Schloss Coswig.
Lit: Kempen, *Simonetti* (1925). Baier-Schröcke, 'Lombardische Stuckateure', p. 113.

SOLDATI, Giovanni Battista (1758–1858)
c. **1790** BIOGGIO, Parish Church, assisted by Giovanni Battista Staffieri (1749–1809).
Lit: Simona, II, p. 15.
A family of the same name was settled at Neggio – Antonio Soldati (d. 1822) worked at Agno; and another group at Porza.

SOLDATI, Tomasso (1665–1743)
A Comasque. Worked in Bohemia.
1694 BROUMOV, Klosterkirche.
c. **1698** PRAGUE, St Ignatius of Loyola.
Lit: Blažiček, 'Lombardische Stuckateure', p. 120.
1704–7 PRAGUE, St Ursula.
Lit: *Arte Lombarda*, 40 (1974), p. 158.

SOLISTA, Michelangelo (*fl.* 1645–1660)
1645–60 SALA CAPRIASCA, S. Antonio. Worked with Andrea Ferrari of Bigorio.
Lit: Simona, II, p. 39.

SOMAZZI, Pietro (*fl.* 1698)
Presumably of the family settled at Montagnola.
c. **1698** TURIN, Palazzo Carignano.
Lit: Carboneri, p. 23, pls. 22–3; Thieme-Becker, 31 (1937); Simona, II, p. 82.

SORRISI, Giovanni Maria (*fl.* 1641–4)
Of Rome. Worked with D. G. Frisone (q.v.) and S. Castellaccio (q.v.).
1641–4 FLORENCE, Palazzo Pitti.
Lit: Campbell, pp. 234–7.

SPINEDI, Francesco (*fl.* 1750)
1750 MENDRISIO, S. Giovanni.
Lit: G. Martinola, 'I Conventi di Mendrisio' in *Bolletino storico della Svizzera Italiana*, Bellinzona 1945, p. 55.

SPORER, Benedikt (*fl.* 1779)
c. **1779** WIBLINGEN, Abbey Church.
Lit: Powell, p. 127.
Thomas Sporer (*fl.* 1743) of Hayd worked in the Feichtmayr circle (q.v.).

STANLEY, Simon Carl (Charles) (1703–61)
Plasterer. Born in Copenhagen, of English father. Trained in Denmark, studied in Amsterdam. Worked 1727 until the summer of 1746 in England, then invited to Denmark by Frederick V to become Court Sculptor, a post he held until his death.
Lit: A. F. Büsching, *Nachrichten von den Künsten*, Copenhagen, III, 1757, pp. 193–200; Ogveke Helsted, *Wellbachs Kunstlerlexi-*

kon, Copenhagen, III, 1952, pp. 262–4, and entries cited therein; Rupert Gunnis, *Dictionary of British Sculptors*, London 1953, pp. 363–6; K. A. Esdaile in *Country Life*, 2 October and 11 December 1937; *Times Literary Supplement*, 3 April 1937.
1721–2 FREDENSBORG, Castle, as apprentice assisted Adam Sturmberg.
c. **1740** LANGLEY PARK, Norfolk.
1744 OXFORD, Radcliffe Camera.
Lit: *Oxford Historical Society*, XIII, 1958.

Attributed work

1728–9 COMPTON PLACE, Eastbourne, Sussex. Stanley was presumably one of four plasterers working under the supervision of John Hughes.
Lit: Beard (1981), pp. 32–3.
c. **1745** KIRTLINGTON, Oxfordshire.
Lit: *Country Life*, 2 October and 11 December 1937.

STILLER, Jakob (*fl.* 1750–3)
1750–3 At SCHONGAU. Assisted F. X. Schmuzer (q.v.).
Lit: Eva Vollmer, *Franz Xaver Schmuzer*, Sigmaringen 1979, pp. 84, 129.

STOCKING, Thomas, senior (1722–1808)
Plasterer who practised in Bristol and the south-west, where his reputation was equal to that of the Rose family (q.v.). His most important commission was at Corsham Court, Wiltshire. Paid £750 between 1763 and 1766, of which £390 was possibly for the very fine Long Gallery ceiling, completed *c.* 1765.
Lit: W. Ison, *Georgian Buildings of Bristol*, 1952, pp. 44–5; Bristol City Archives, Apprentices Book, 1764–77.

TADDEI, Carlo Giuseppe (1702–70)
Born at Gandria, son of Michelangelo Taddei and Caterina Giovanni Serena of Arogno.
Lit: Brentani, IV, p. 394.

TADDEI, Michel Angelo and **Francesco Antonio** (*fl.* 1777)
Presumably of a younger generation than the Michelangelo Taddei noted above.
1777 AUGUSTENBORG (Denmark), Castle.
Lit: Bredo L. Grandjean, pl. 206.

ÜBELHERR (**Ubelhor**), **Johann Georg** (1700–63)
Of Wessobrunn. Son-in-law of Joseph Schmuzer (1683–1752), through marriage to his daughter Maria Agatha in 1741. Assisted the Schmuzer family (q.v.) at several commissions.
1734 WILHERING, Abbey Church. Worked with Holzinger and Feichtmayr.
c. **1738** DIESSEN, Parish Church. Worked with F. X. Feichtmayr.
1748 KEMPTEN, Residenz. Worked with F. X. Feichtmayr and Johann Schütz.
Lit: Norbert Lieb, *Rokoko in der Residenz von Kempten*, Kempten 1958.
1745–51 ETTAL, Abbey Church. Worked with Joseph Schmuzer (q.v.).
Lit: Lieb (I), pl. 139.
c. **1750** AMORBACH, altars, assisted J. M. Feichtmayr.
1751–5 ANDECHS, Priory Church.
1764 VIERZEHNHEILIGEN, Pilgrimage Church. Stuccoes completed after Ubelherr's and F. X. Feichtmayr's death by J. M. Feichtmayr.

VASSALLI, Francesco (*fl.* 1724–63)
Of a family long settled at Riva St Vitale in the Ticino. Although little is known of him, he was working in England by 1714. He may have joined the Scottish team of Thomas Clayton for a short time in the 1740s. He had a foreman, John Johnson, in 1755. His work in England included:
1715 DUNCOMBE PARK, Yorkshire.
1724 SUTTON SCARSDALE, Derbyshire.
1725 DITCHLEY, Oxfordshire.
1730 ASKE HALL, Richmond, Yorkshire.
1730–1 TOWNELEY HALL, Burnley, Lancashire.
1732–3 PARLINGTON HALL, Yorkshire (dest.).
1736–7 CASTLE HOWARD, Yorkshire, Temple of the Four Winds.
1751–2 TRENTHAM, Staffordshire (dest.)
1753 PETWORTH, Sussex.
1758 HAGLEY HALL, Worcestershire, Hall, signed panel, Dining Room, Long Gallery.
1759–61 CROOME COURT, Worcestershire.
1763 SHUGBOROUGH, Staffordshire, Dining Room, Library.
Lit: Beard (1981), Dictionary entry gives attributed work.

VERGA, Francesco (*fl.* 1700)
Of Mendrisio.
1700 LIGORNETTO, Parish Church, S. Lorenzo.
Lit: G. Martinola, 'La chiesa di Ligornetto' in *Bolletino storico della Svizzera Italiana*, Bellinzona 1946, p. 212.

VERHELST, Aegidius, the elder (1696–1749)
1749 DIE WIES, Pilgrimage Church, high altar, figures of the Four Evangelists. Assisted the Zimmermann brothers (q.v.).
Lit: Lieb (I), pl. 157.

VOLPINI, Giuseppe (*fl.* 1726)
1726 MUNICH, Schloss Nymphenburg, Chapel, figure of Mary Magdalene.
Lit: L. Hager and H. Kreisel, *Nymphenburg*, Munich 1938, p. 56.

WALNEGRA, Simone (*fl.* 1st half 18th c.)
From the Intelvi valley.
1726 BLANKENBURG, Castle. Worked with Michele Caminada and Carlo Rossi.
Lit: Amerio, p. 99.

ZAMMELS, Burkard (1690–1757)
1718–19 POMMERSFELDEN, Schloss Weissenstein, staircase, figures of Juno, Venus, etc.
Lit: H. Kreisel, *Das Schloss zu Pommersfelden*, Nuremberg 1968.

ZIMMERMANN brothers, Johann Baptist (1680–1758) and **Dominikus** (1685–1766)
Of Wessobrunn. Sons of stuccoist Elias Zimmermann (1656–95).
1709–10 EDELSTETTEN, Frauenstift Church.
Lit: Hitchcock, *G. R.*, pl. 1; Ernst Gall, *Handbuch der deutschen Kunstdenkmäler, Östliches Schwaben*, Munich 1954, p. 138.
1711–13 BUXHEIM, Abbey Church, stucco surround (D.Z.) of frescoes (J.B.Z.).
Lit: F. Arens and F. Stöhlker, *Die Kartause Buxheim*, Buxheim 1962; Hitchcock, *G. R.*, pls. 2–3.
1716–18 OTTOBEUREN, Abbey, Library. (J.B.Z.)
Lit: Lieb, *Ottobeuren*, Augsburg 1933.
1716–18 MÖDINGEN, Choir arch (D.Z.); frescoes 1719, 1722 (J.B.Z.).
Lit: J. Schöttl, *Kloster Maria-Mödingen*, Munich 1961.
1719–20 LANDSBERG-AM-LECH, Rathaus façade stucco. (D.Z.)
1724–5 SCHLEISSHEIM, Great Hall, Kammerkapelle.
Lit: Hitchcock, *RASG*, p. 41.
1724–5 BENEDIKTBEUERN, Abbey, Library.
Lit: K. Mindera, *Benediktbeuern*, Munich 1957.
1726–7 BUXHEIM, Parish Church. (D.Z.)
Lit: F. Arens and F. Stöhlker, *Die Kartause Buxheim*, Buxheim 1962.
1728–30 STEINHAUSEN, Pilgrimage Church, figures, balustrade, etc. (D.Z.); frescoes (J.B.Z.).
Lit: Hitchcock, *G. R.*, pl. 20.
1729 WEYARN, Abbey Church.
Lit: Hitchcock, *G. R.*, pl. 21.
1731–2 BENEDIKTBEUERN, Abbey, west wing, Neuer Festsaal. (J.B.Z. and his son Joseph)
Lit: Mindera, pp. 33–5.
1734–9 NYMPHENBURG, Amalienburg, Spiegelsaal and other rooms. (J.B.Z.)
1738 PRIEN-AM-CHIEMSEE, Parish Church, angels, *stuckmarmor* pulpit.
c. **1743** MUNICH, Berg-am-Laim, St Michael.
Lit: Lieb (I), pl. 83.
1749 DIE WIES, Pilgrimage Church, with J. G. Übelherr, A. Verhelst the elder, and A. Sturm.
1751–5 ANDECHS, Pilgrimage Church.
Lit: R. Banerreis and H. Schnell, *Die Heilige Berg Andechs*, Munich 1955.
1754–6 SCHÄFTLARN, Abbey Church.
1755–7 MUNICH, Schloss Nymphenburg, Festsaal.
Lit: L. Hager, *Nymphenburg*, Munich 1938, pp. 57–60.

ZÖPF, Thassilo (1723–1807)
Married Maria Elisabeth Übelherr.
1757–9 WESSOBRUNN, Parish Church.
1760 FÜRSTENFELDBRUCK, Abbey Church.
1764, 1766–7 POLLING, Abbey Church.
1771 WESSOBRUNN, Kreuzbergkapelle.
1774 At MOORENWEIS.
1780 BEUERBERG, Marienkirche.
Lit: Eva Vollmer, *Franz Xaver Schmuzer*, Sigmaringen 1979, p. 89, pls. 132–3; D. Pflüger, 'Der Wessobrunner Stukkator, Thassilo Zöpf', Dissertation, University of Munich (1973).

ZUCCOLI, – (*fl.* late 17th c.)
1684 VIRA E MEZZOVICO, Altar of the Madonna bears inscription 'Arrigoni e Zuccoli fecero'.
Lit: Simona, II, p. 11.

Indexes of Persons and Places

References to the illustration-captions are indicated by plate numbers in italic type for monochrome, and by roman numerals for colour

Index of Persons

Index of Places

Photographic acknowledgments

Roman numerals refer to colour plates, arabic numbers to black and white illustrations

Alinari 4, 10, 22, 25, 27, 28, 30, 31, 32, 35, 40
Anderson 6, 9, 11, 15, 26, 33
B. T. Batsford 67
Geoffrey Beard II, XII, fig. 5
T. J. Benton 39
Osvaldo Böhm 14
Nationalmuseum, Copenhagen 120, 121
Country Life 20, 66, 68, 112
Courtauld Institute of Art, London 7, 12, 24, 55
Irish Tourist Board, Dublin 116, 117, 127
Giraudon IV, 16
Hirmer Fotoarchiv 76, 104
A. F. Kersting V, VI, VII, VIII, XIV, XV, XVI, 50, 53, 54, 56, 57, 59, 73, 74, 75, 77, 80, 81, 85, 86, 87, 88, 89, 90, 91, 98, 99, 100, 108, 110, 126, 130
British Museum, London fig. 8
Bildarchiv Foto Marburg 8, 29, 83, 94, 134
Mas 69
Metropolitan Museum of Art, New York, Rogers Fund, 1906: 34, 111
George Mott 115
Stadtmuseum, Munich 115
National Monuments Record 106, 109, 129
Nonsuch Excavation Committee 18, 19
Ashmolean Museum, Oxford fig. 1, fig. 13
Archives Photographiques, Paris 42, 128
U. Pfistermeister IX, X, XI, XIII, 45, 46, 47, 48, 49, 58, 78, 79, 82, 92, 93, 101, 102, 103
Polish Academy of Sciences, Institute of Art, Photographic Library 64
Presse-Bild-Poss 51, 52
Státní Ústav Památkové Péče a Ochrany Přírody, Prague 21, 62, 63
Fototeca Unione, Rome 1, 2
Gabinetto Fotografico Nazionale, Rome 107
Royal Commission on Ancient Monuments, Scotland 113, 114, 132
Scala I, III, 36, 37, 38
Helga Schmidt-Glassner 43, 60, 61, 72
Edwin Smith 13, 17, 41, 65, 70, 71, 95, 96, 97, 105, 118, 119, 122, 123, 124, 125, 131, 133
Nationalmuseum, Stockholm fig. 10
Bundesdenkmalamt, Vienna 44
Stadtisches Museum, Weilheim 84
Anton Fuchs, Würzburg fig. 6, fig. 7

Reproductions from books:

Jean Berain, *Ornamens inventez par J Berain*, Paris 1711: fig. 11
Johann Jacob Billers, *Neües zierathen Buch von Schlingen und Bändelwerk*, Augsburg 1710: fig. 15
François Cuvilliés, *Livre de Portion de Plafonds et d'un Poëlle*, Paris 1738: fig 14
Paul Decker, *Fürstlicher Baumeister oder Architectura civilis*, vol 1, Augsburg 1711: fig. 12
Joseph Moxon, *Mechanick Exercises*, London 1703: fig. 3
P. N. Sprengel, *Handwerke und Künste in Tabellen*, Berlin 1772: fig. 4